The Space Elevator Concep

The concept of a Space Elevator started over a century ago but started to move from science fiction to reality in the 1990s with the promising advent of a super strong material, carbon nanotubes, that theoretically had the strength to build a ribbon into space.

NASA started investigating the concept which led to the publication of the first detailed analysis of whether it was possible, written by Dr Brad Edwards, in a NIAC paper.

Early excitement of an immediate breakthrough, maybe even constructing one by 2010, gave way to the reality of producing carbon nanotubes of the required strength, outside the laboratory, in commercial quantities.

As I write, in 2020, we are still awaiting this breakthrough and the project has moved back to the late 20s, maybe the 2030s or 2040s.

But it will happen, and when it does, it will replace rockets as the easier and cheaper way to leave Earth, opening up space travel the way airplanes opened up world travel.
This series examines the technical aspects and how it will be deployed.

The first book in the Space Elevator 2020 series covers the history of this exciting project.

Contents

1. REFLECTION

For as long as our ancestors looked up into the heavens, they must have speculated about ways of reaching them. The idea of a rocket, to travel into space, was first conceived a century before technology had developed enough to make it happen. The same is true for the space elevator, essentially a rope, albeit a very strong one, stretching into space, on which cable cars can climb, which idea came to Konstantin Tsiolkovski in his 1895 book.

The post-war generation has seen new ideas, concepts and technology appear at the fastest rate in history. The world of 2018 bears no relation to the world of the 1950's.

In 1957...
no rocket had ever traveled to space.

Yet 12 years later in 1969...
astronauts were walking on the Moon!

***

In 1903, no airplane had ever flown.

Yet, in 2020 we routinely fly around the world, to London for business, to Hawaii for holidays.

***

In the last half of the 20th century, huge, expensive rockets were the only way to travel to space.

By the 2030's or 2040's, traveling to space and the Moon by space elevator will be the norm.

***

The name of Dr. Bradley C. Edwards, recipient of the Arthur C. Clarke Innovator Award in 2005. has been associated with the Space Elevator, since his pioneering work with NASA's NIAC (NASA Institute of Advanced Concepts) on the project in the 1990's.

The project was explained in detail in the 2006 book Leaving The Planet By Space Elevator (Edwards, Ragan). With the passing of Ragan, Phillips has taken over the role of refreshing that work, addressing the current status of the project.

This book is part of a series analyzing the Space Elevator concept in more detail, published in 2020.

2. IT'S ROCKET SCIENCE, BUT NOT AS WE KNOW IT

Right now, to travel into space, we use rockets. As space travel vehicles, they only have a half-century history, the first launch being back in 1957. Retired baby-boomers have seen the entire history of space travel to date!

They get there, for sure, but they are expensive, and in terms of space travel, quite slow. The speed achieved by rockets to achieve escape velocity from Earth, about 11.2 km per second (kph), looks fast to us. Conventional rockets travel in space at about 28,000 kph.

But space is big, very big. In the words of Douglas Adams (Hitch-hikers Guide to the Galaxy)

"Space is big. Really big. You just won't believe how vastly, hugely, mind-bogglingly big it is. I mean, you may think it's a long way down the road to the chemist [drugstore], but that's just peanuts to space."

Even at that speed, and optimal conditions, a spacecraft takes this long to travel to our closest destinations:

Moon - about 4 days
Venus - 3 months or more
Mars - anywhere from 4 months to 12 months

Jupiter - 4 years
Saturn - 5 years
Uranus - 6 years
Neptune - 8 years
Pluto - 8 to 10 years

These figures are minimums depending on how the planets are lined up. They orbit the sun at different speeds and your destination planet needs to be aligned, in the right place, to commence a journey.

Orbit periods (in Earth years)

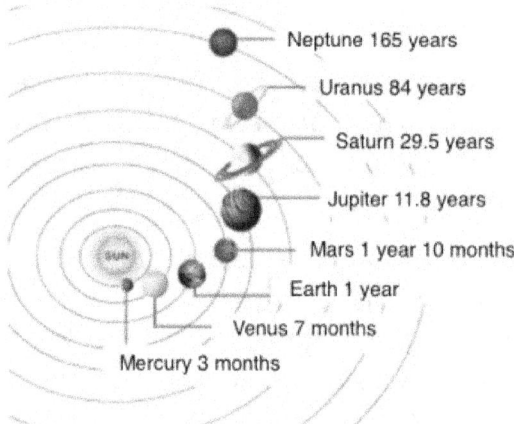

Neptune 165 years
Uranus 84 years
Saturn 29.5 years
Jupiter 11.8 years
Mars 1 year 10 months
Earth 1 year
Venus 7 months
Mercury 3 months

The various orbital periods underline the distances involved in exploring our solar system. Mars takes nearly two of our years to orbit the Sun. But, go to Jupiter, the next planet out, and the orbital time jumps dramatically to 11.8 years, nearly six times as long.

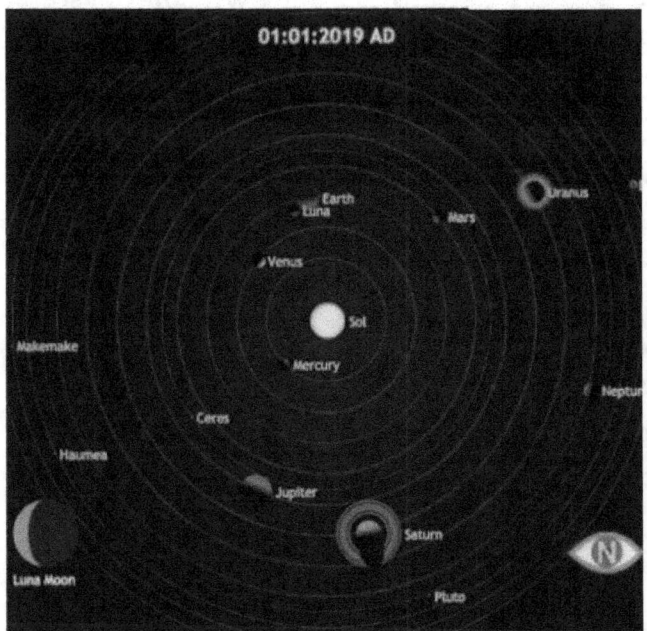

Location of the planets at the beginning of 2019
(distance from Sun not to scale)
Source The Planets Today

At the beginning of 2019, Mars was further away from Earth than Mercury, while Jupiter and Saturn were on the other side of the Sun.

The nearest star system around Alpha Centauri? Forget it. It's 4.4 light-years away, about 26 trillion miles, and a conventional rocket would, at best, take 165,000 years to get there, almost as long as the entire history of Home Sapiens to date.

So, we're struggling to even get around our own solar system, never mind leave it, but just in case someone invents warp drive, these are our nearest neighbors to visit:

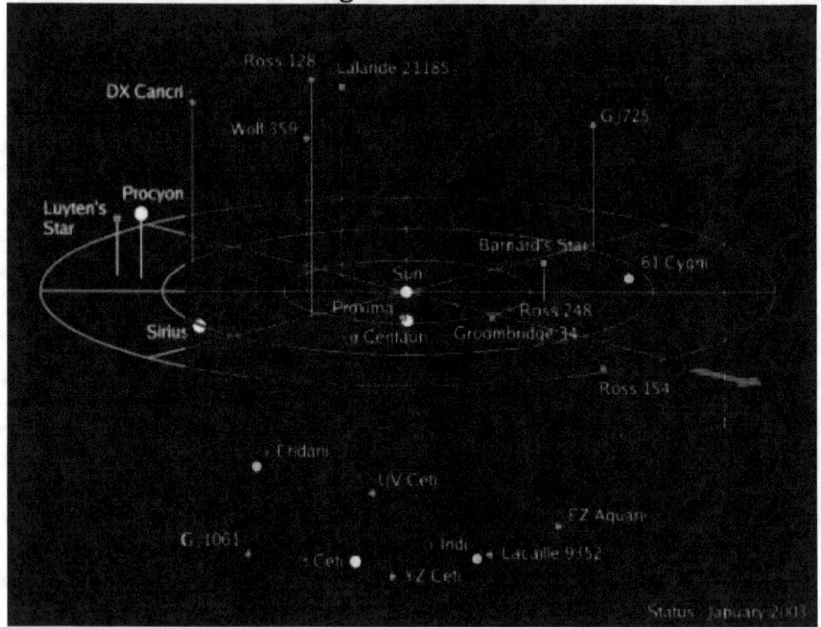

Stars near us

We have a couple of dozen stellar neighbors within ten light-years, most of which are visible at night. In the southern sky, I have a fine view of the Centauri system, in the pointers to the Southern Cross, while Sirius, the Dog-Star, is the brightest star in the sky, near the constellation of Orion.

Most of the cost, and energy, involved in space travel is simply to get off of the Earth. With our heavy gravity well, leaving the planet is expensive in cost and energy. Once you are well away from Earth, in orbit around the Moon, say, travel becomes relatively easier.

So, what's on the drawing board to make space travel cheaper and easier? At this point, rocket scientists look down at their feet, get embarrassed and change the subject.

In the realms of science fiction, many ways of traveling through space and time have been imagined. Let's not knock them, for even rockets were once science fiction, and many stories have had a way of becoming science fact, eventually. But until someone figures out how to create the warp drive of Star Trek, or build wormholes in space, the only other concept around which will take us beyond rockets is the Space Elevator.

With space elevators in use, on Earth, our Moon, Mars and maybe elsewhere, the space game changes. The cost of getting off of Earth falls to about 5% of the current cost. Instead of infrequent rocket launches (in 2017 the number of rocket launches were just 68 in total: 29 (USA), 20 (Russia), 19 (China)), the advent of space elevators would see cable cars riding into space dozens of times a day, making the space travel experience more akin to air travel. The cost of lifting fuel, water and supplies falls, enabling interplanetary rockets (yes, they'd still be needed for services between the Moon, Earth and planets) to travel faster by several multiples. We can place space stations in orbit around the Moon and Mars, for starters, and service them with daily scheduled cargo crafts. Crewed flights to the Moon and Mars will be safer and better supported by a network of space stations and cargo vehicles.

In short, to build colonies in our solar system, to mine planets and asteroids, to facilitate human space travel, isn't really feasible until we have space elevators in place. Doing it by rocket is the hard way, equivalent to traveling the world by hot-air balloon instead of airplanes.

Will space elevators be the pinnacle of space travel technology? It isn't a silly question. If we cannot dream up anything better, no warp drive, no wormholes, then, in practice, we can populate our solar system, but nowhere else, no interstellar space travel, no checking out Proxima Centauri or other habitable worlds.

We can imagine accelerating small satellites to a sizable fraction of light speed, sending a satellite to Proxima Centauri, say, but we are talking a time scale of decades and centuries just to do this.

Even if such a satellite detects a pristine habitable world, the sort we'd all like to explore, it would be a tough project of a thousand years or more to send people there.

Maybe the space elevator is the pinnacle of space technology. This would be consistent with what we see in our galaxy and our universe to date. Everything we see in the universe can be explained by natural forces. To date, we see nothing needing an advanced technology or civilization to explain it. Innately we believe there must be intelligent life out there in the universe, but if life is, in practice, limited to its own solar system, then concerns about being invaded by Klingons are overrated.

So, in this series, we paint a picture of life by the end of the 21st century where, by 2100, we have colonized every conceivable niche of our solar system, where space travel in 2100 is as routine and commonplace as air travel on Earth is today. The key to it all? The Space Elevator, as the next technology to replace rockets for leaving the planet.

3. ORIGIN OF THE SPACE ELEVATOR CONCEPT

So many ideas have started their life as science fiction. In fact, long before the term "science fiction" was conjured up, visionary writers of the 19th century were dreaming of a future that accurately foresaw coming discoveries in many aspects. (Science fiction can be described as dreaming the future in detail!)

The 19th century was a continuation of the tremendous burst in human thought and endeavor known as the "enlightenment". New discoveries seemed to come one after the other in quick succession: magnetism, electricity, gas mains, electric lights, telescopes, microscopes, clocks, the telephone, the internal combustion engine, trains and cars.

People came to expect ever more progress, just like the world today.

Famous writers of the day, including Jules Verne, William Morris and H.G. Wells, used their imagination to envisage airplanes, rockets, time-travel, and journeys to the center of the Earth and to the Moon. Other scientists had worked out the math of rockets and travel to the Moon and planets, long before the first rocket was ever invented. Those science fiction writers had thought through most of the problems of being in space, long before anyone actually went there. These ideas, musings and writings had the real potential to inspire researchers and engineers to reach for the heavens.

Science fiction has a way of turning into science fact. This is likely to be the case with the alternatives to rocket science also.

In terms of far out ideas for space travel, most people would imagine some kind of warp drive such as used in the Star Trek episodes, as being the ideal way of moving around the universe. Great idea, but no one seriously imagines creating such a warp drive anytime soon.

The Holy Grail of launching into space would surely be some kind of antigravity engine. If only we could neutralize gravity and simply float into space, unbound by the deep gravity well of our Earth. One day this may happen, but we don't see it on the horizon. Scientists are still struggling with the nature of gravity and speculation as to the forces that control it. We are a long way from creating any kind of antigravity device. So, for now, this must remain in the realm of science fiction.

Tsiolkovski

Konstantin Tsiolkovski (orTsiolkovsky) is credited with having first taken the intellectual leap, imagining such a tall tower into space with his 1895 paper "Day-Dreams of Heaven and Earth", as reported on by Jerome Pearson.

Tsiolkovski was a Russian schoolteacher in St. Petersburg, and he dreamt up a world of tall towers and cosmic railways reaching into space. It was quickly apparent that known materials would not be strong enough to carry their own weight for such a height and this led Tsiolkovski to conclude that such a tower was not possible.

Along the way he also discovered the concept of geostationary orbit, and he calculated the radius of such orbits for Earth and other planets.a 1997 article.

Having written in Russian, the work of Tsiolkovski was slow to reach the Western World, but meanwhile attention became focused, from 1905, on the design and construction of airplanes, and from the 1930's, on the development of rockets, first for military use and later for space travel.

Some writers conjured up the building of incredibly tall towers from Earth as a way of reducing the weight of the suspended cable. However, no matter how it was done, the combined weight was too much to support itself, never mind any cargo. Even with the best of materials available in the 20th century, it is theoretically possible to build structures only a kilometer or so high. The 828 meter tall Burj Khalifa in Dubai is the worlds tallest building but this is still a long way short of the intended 100,000 km cable length.

1986 CCCP stamp showing Tsiolkovski

Image extract from the Pravda article

The strength-of-material issue seemed to be a killer, and Space Elevators looked destined to remain science fiction. Since there was no known material that could support its own weight over such a length, the whole concept was deemed impossible from the start.

Artsunov

The idea was not revisited until 1960 when another Russian, Leningrad engineer Yuri Artsutanov, set out a realistic structure for a Space Elevator, in a Pravda Sunday Supplement, "To the Cosmos by Electric Train" published by Pravda in 1960. (See here for article links and the full article.)

In the article, he introduces the idea with:

"We wish to propose one more design for such a station, one directly connected to Earth. The realization of this design may make the trip into cosmic space only a bit more complicated than a trip today from Moscow to the suburb of Mozhaika on an electric train...

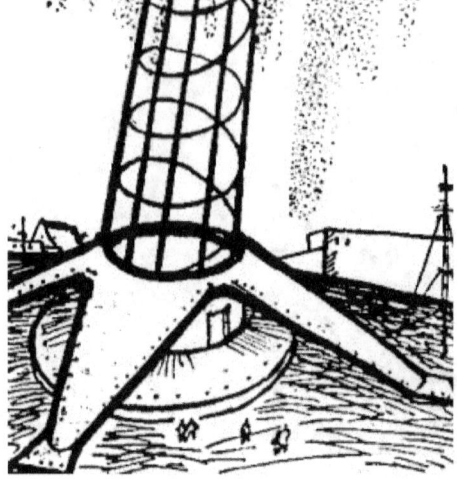

Take a little piece of string and attach to it a stone. Begin to rotate this primitive sling. Under the influence of centrifugal force the stone will try to pull itself away and tightly stretch the rope. Well, what will happen if one fastens such a "rope" to the Earth's equator and, having flung it far into the cosmos, one hangs on it an appropriate load?

Calculations show (any student of the upper grades of middle school can work them out) that if the "rope" is sufficiently long, then centrifugal force will also pull it out, not letting it fall to Earth, just like the stone stretches out our string.

Indeed, the Earth's force of attraction lessens in proportion to the square of the distance, and centrifugal force grows with the increase in distance. And already at a distance of about 42 thousand kilometers centrifugal force turns out to be equal to the force of gravity."

Assuming he meant 42,000 km from the center of the planet, that figure is quite accurate, since what we now call geostationary Earth orbit (GEO) is 35,500 km from sea level, 42,200 km from the center of the Earth.

Isaacs, Vine, Bradner and Bacchus

In 1966, John Isaacs along with Vine, Bradner and Bacchus of the Scripps Institute of Oceanography, La Jolla, California rediscovered the concept, though they were probably not aware of the Russian works. They proposed a very thin wire but assumed that it would get severed by meteors and wouldn't last. Given the assumptions of materials of the day, a thin wire would still not be strong enough to support its own weight.

It may seem strange for oceanographers to be thinking about a Space Elevator, but they were a group of people concerned with the deployment of very long cables underwater, so perhaps it was not so surprising that they considered also the prospect of a cable going in the other direction.

Jerome Pearson

In 1975, Jerome Pearson published a paper on an orbital tower. Unaware of the earlier works he claimed credit for the idea of a Space Elevator that was extended from his orbital tower ideas. He did some good basic work on a Space Elevator concept. In his paper "the theoretical possibility is examined of constructing a tower to connect a geostationary satellite to the ground."

Acknowledging the difficulty of finding a material strong enough to build it with, he was first to put into print the novel ideas of "tapering the cross-sectional area of the tower" and using "perfect-crystal whiskers of graphite" as the building material. He was on the right track with the graphite crystals but it was two decades later before the carbon nanotube was discovered.

He also considered a "lunar anchored satellite", that is, an Elevator on the Moon. In 1980, he wrote a related article on asteroid retrieval using similar principles. In later years Pearson continued to work on a lunar elevator design.

"Pearson, unaware of the earlier work, independently discovered the concept, but published it in the international journal Acta Astronautica, and thus gave the space elevator to the international aerospace community for the first time. His discovery included using the space elevator for zero-net-energy space launching, and for launching payloads from the elevator tip to reach other planets without requiring rockets. He also was first examine the dynamics of actually lifting payloads up the elevator, and found limitations on the speeds of ascent, akin to the critical velocities of a rotating shaft and the periodic loads from soldiers marching on a bridge." See article.

Pearson wrote invited articles on space elevators and tethers for Encyclopædia Britannica and New Scientist, and was featured discussing space elevators in the Discovery Channel series "Science of the Impossible".

4. IT ENTERS SCIENCE FICTION

Sir Arthur C. Clarke

Pearson briefed the science fiction writer Sir Arthur C. Clarke on the concept, which was used in his fictional work The Fountains of Paradise in 1979. The book postulated building a tower up from Earth to meet a cable hanging down from space. His book had the tower rising from a fictional island on the Equator, based on his home island of Sri Lanka, or Ceylon as it used to be known.

Sir Arthur C. Clarke, Photo courtesy of the Office of Sir Arthur

It was the first work of fiction based on the space elevator idea. As such, the technology imagined in the story was primitive, out of date only twenty years later. But its influence was such as to be the basis of prevailing misconceptions, even today. To this day, there are reviewers who believe that the Space Elevator concept requires a 50 km high building to be erected first! It is an indication of how fiction gets more attention than fact.

In an address to the 30th International Astronautical Congress at Munich on 20 September 1979, Sir Arthur gave a presentation on the practicalities of the Space Elevator concept, probably the first such public briefing of the idea. It was printed in the journal Advances in Earth Oriented Applied Space Technologies, Volume 1 in 1981.

In that address he could already refer to "the rapidly expanding literature of the subject". Acknowledging the many that had thought of the concept, Sir Arthur referred to related work in 1969 by Collar and Flower, and said:

"Collar and Flower did mention that it would be possible for the cable to reach all the way down to the Earth's surface, though they did not elaborate on this point, and were apparently unaware of earlier work in this field. For it now appears that at least half a dozen people invented the Space Elevator quite independently of each other, and doubtless more pioneers will emerge from time to time."

More science fiction

Three of the greatest science fiction writers are Sir Arthur C. Clarke, Kim Stanley Robinson and Stephen Baxter, and all three showed their ability to foresee the future by incorporating the Space Elevator concept into their books.

Those books were, respectively The Fountains of Paradise 1979, Red Mars 1992 and Sunstorm 2005, the latter being co-authored by Sir Arthur C. Clarke and Stephen Baxter.

In 1992 Kim Stanley Robinson wrote Red Mars one of a trilogy of books centered on the colonization of Mars. A key aspect of the story was the use of a Space Elevator constructed on Mars. Robinson envisaged a metal cable, which however would not be feasible even given the low gravity of Mars, but with hindsight he had a key fact correct: the cable was hanging down from space, not built up from the ground. Also, the planet had only one cable, not several, so the loss of the single cable would be devastating which it was in several different ways integral to the plot.

As part of the plot, and central to an interplanetary battle, the inhabitants of Mars at one point disconnect their Elevator cable and watch as it collapses back onto the planet. The Earth invaders no longer have rockets that can descend to the planet, and so the Martians declare their independence from Earth.

The book Sunstorm by Sir Arthur C. Clarke and Stephen Baxter links to the authors of Leaving The Planet By Space Elevator, since the inclusion of the Space Elevator was based on the technology, and the Earth base of Perth, Australia, as set out by the authors. So Sunstorm was the first scifi novel with a sound technical basis.

Sunstorm novel, front cover

In the 21st century, in scifi, the space elevator has matured from "impossible science" into an established concept and has appeared in many scifi novels. Once a fanciful idea, it is now considered possible, and has taken its place in the pantheon of future space exploration.

Wikipedia lists over 56 sci-fi books to date including a space elevator as part of the story. (https://en.wikipedia.org/wiki/Space_elevators_in_fiction)

Sir Arthur C. Clarke also wrote 2061: Odyssey Three, in 1987, featuring a space elevator again, while Kim Stanley Robinson wrote 2312, written in 2012, featuring 37 space elevators connected to Earth. Edwards, co-author of this book, edited a book of short stories based on elevators called Running The Line in 2006.

The effect of all these books, while still fiction, has been to familiarize readers at large with the concept.

Back to reality

With the idea having been published in English, other writers added different angles to the idea, including Hans Moravec, Paul Birch, Keith Lofstrom, Rod Hyde, and Paul Penzo. These ideas included space tethers, i.e. long cables that rotated in space without being attached to any planet or Moon.

But what good is an idea if it can never turn into reality?

This, then, was the situation: mankind was desperate to leave the planet. We had rockets and we proved that we could physically go into space and land on the Moon. But that was extremely expensive.

Meanwhile, the idea of a ladder into space, a Space Elevator or similar, remained just that: an idea. All known materials have a particular tensile strength: stretch them beyond that and they break. A 100,000 km long cable could not be produced and put in place. Quite simply, it would break under its own weight.

So that was the first century of the Space Elevator concept. There was always the hope that, one day, a stronger material would be discovered. Both Artsutanov and Pearson had speculated that some form of carbon (carbon "whiskers") might be such a material, but none had been created.

Sir Arthur C. Clarke was once asked: "When will a space elevator be built?" His flippant reply was: "About 50 years after everyone stops laughing." In 1979, when he wrote his sci-fi book, that seemed fair enough. But science progresses rapidly, and a couple of decades later, the game changed and the idea started to be taken seriously.

5. THE BREAKTHROUGH DISCOVERY OF THE 1990S

In the 1990s a new discovery was made that would change everything.

A new form of carbon was discovered known as carbon nanotube - a close relative to the fullerenes or "buckyballs". This form of carbon might be woven into cables or sheets although, as we write, this is still only being achieved in the laboratory.

However, the exciting thing about this material is that carbon nanotubes have been measured at strengths 400 times as strong as steel (pound for pound). This makes it potentially the strongest material known in the universe. We need roughly 100+ times the strength of steel to build the Space Elevator.

The late Richard Smalley (who won the 1996 Nobel Prize in chemistry for the discovery of fullerenes) along with Robert F. Curl, Jr., of Rice University, Houston, and Sir Harold W. Kroto of Great Britain were the joint discoverers of carbon buckyballs, a hitherto unknown allotrope of carbon that consists of 60 carbon atoms in the shape of a soccer ball.

Whilst scientists and industry were excited about carbon buckyballs for a host of potential applications, the late Richard Smalley and Boris Yakobson at North Carolina State University wrote of their potential for a space cable.

In an article now posted online in 1997 they reported on their experimental production of carbon nanotubes as follows:

"In computer simulation tests, Boris Yakobson and his colleagues at North Carolina State University subjected the stringlike nanotubes to extreme stresses." We wanted to see how flexible or brittle they are," said Yakobson, "and found they're amazingly flexible, unusual for graphite, which in large scale seems brittle."

In twisting, bending, compressing and putting the virtual nanotubes on the rack, the physicists showed the structures could be stretched by almost 30 percent without breaking. Theoretically, they were strong enough to create a thread able to sustain 150 GPa (gigapascals), equivalent to holding a 20-ton weight on a 1 millimeter thread.

Hearing of the results, Richard Smalley communicated with Yakobson. Smalley suggested nanotubes might be the real material to build the Space Elevator that Arthur C. Clarke describes in his novel Fountains of Paradise -- a 23,000 mile [35,000 km] cable from a space station to Earth. No known steel or other material could avoid snapping under its own weight at that length - until nanotubes. Only the tiny fullerene strands might sustain their weight when spanning 23,000 miles."

Smalley realized the critical implications of the strength of carbon nanotubes for a Space Elevator, though he assumed that it would stretch only to GEO, not beyond as it is now planned.
NASA was briefed on the idea, and convened a workshop to consider a Space Elevator, the first recorded serious analysis of the proposition.

The workshop titled "An Advanced Earth-Space Infrastructure for the New Millennium" was held in June 1999 at Marshall Space Flight Center, with the conference managed, and a report compiled by, David Smitherman. (Smitherman, 2000).

Attendees at that conference included names already mentioned such as Jerome Pearson (Star Technology and Research Inc.), Paul Penzo (Jet Propulsion Laboratory), Richard Smalley and Ken Smith (Rice University) and others.

The conference report was in favor of further analysis of the Space Elevator concept and viewed it as something that would possibly be built towards the end of the 21st century. (Though since then, the timetable has moved forward to the 2030s or so.)

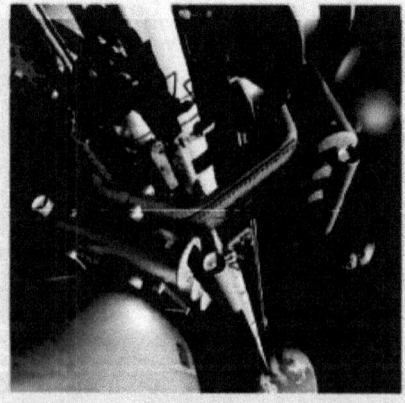

Space Elevators

An Advanced Earth-Space Infrastructure for the New Millennium

Compiled by D. V. Smitherman, Jr.
for National Aeronautics and Space Administration

6. Dr. Bradley Edwards and NIAC

Los Alamos National Laboratory (LANL) physicist Dr. Bradley Edwards, co-author of the subsequent "Leaving the Planet by Space Elevator" book with Philip Ragan, was also working on the same concept.

He started work on the Elevator idea at the end of 1997 and submitted a paper on it to Acta Astronautica in early 1999. The paper took about 18 months to be published (Volume 47, 2000) and displayed many of the themes and conclusions that appeared in his later NIAC Stage 1 report.

Dr. Edwards was not an attendee at the Marshall conference but was already pursuing the concept. He completed the NIAC Stage 1 report while still at LANL but then left for NIAC to concentrate on the Stage 2 report.

His conclusions were different from the workshop conclusions in many material respects, in particular:

* Ditching the workshop concentration on tall towers in favor of a ribbon (as opposed to a round cable) that dropped all the way to Earth.
* In focusing entirely on the use of carbon nanotube for the ribbon.
* In avoiding the use of an asteroid for the counterweight.
* In designing the minimal useable system with current technology where possible.
* In using the Elevator to build itself.
* In the use of laser power beaming.
* In several specific solutions to survival issues.
* In adopting a 100,000 km ribbon length instead of a cable to GEO only, and in avoiding the idea of interim tethers in favor of going straight to the building of the entire ribbon.

He decided to carry out some rigorous research into the question of whether it really was practical to use this material for a Space Elevator ribbon.

Dr. Bradley C. Edwards

In his first groundbreaking piece of research for the NASA Institute for Advanced Concepts (NIAC), Dr. Edwards demonstrated in detail that it was possible. Ever since, work has been underway to prove up the concept with the aim of launching the first Space Elevator in the 2030s in time to support the Mars missions.

That item of research was more rigorous than anything done previously. Conducted in 1999 and presented to NIAC in June 2000, he established that the carbon nanotube (CNT) material had the required strength, and he addressed the physics of deploying a ribbon into space, demonstrating that it was possible.

(Note: later work on the concept by NIAC ended when NIAC was shut down in 2007, the work passing into private companies.)

However, he was soon to discover that it is not easy to seriously promote what had been only science fiction. In the introduction to his pioneering study for NIAC, he wrote:

"This manuscript is the result of a six-month investigation I conducted for NASA under the NASA Institute for Advanced Concepts (NIAC) program. Even though this is the final report for that study, it is really just the beginning.

The study had the same simple title as this manuscript, The Space Elevator. The study itself was far from simple however. The object was to investigate all aspects of the construction and operation of a Space Elevator, a concept that up until this time had been confined to the realm of science fiction."

Not only did he analyze whether it was theoretically possible but he also addressed the practical issues that would have to be faced in building the first Elevator.

So by the year 2000, it had been demonstrated that the material was theoretically capable of carrying its own weight for a 100,000 km ribbon, plus the weight of cargo ships crawling up and down it, with a considerable safety margin.

A second stage of research followed, with the Phase 2 report to NIAC presented in 2003. This phase was expanded and included contributions from some 20 institutions and 50 participants. It focused on the laboratory production and testing of carbon nanotubes, as explained in the summary extracted from the paper:

"Currently NASA and all space agencies are completely dependent on rockets to get into space. Several advanced propulsion systems are being examined by NASA and others, but few, if any, of these technologies, even if perfected, can provide the high-volume, low-cost transportation system that will be required for the future space activities mankind hopes for. A system that may have the required traits is the one that we examined in our Phase I [paper], the Space Elevator. The Space Elevator, a cable that can be ascended from Earth to space, is unlike any other transportation system for getting into space.'

'Our Phase I [paper] laid down the technical groundwork examining all aspects of a proposed first elevator, but was unable to test many of the designs and scenarios proposed. The hurdles were found and the technology requirements for the system quantified.'

'Even we, the proposers, were surprised by the apparent feasibility of the Space Elevator, the availability of almost all of the required technology, and the affordability of the first elevator.'

'Our Phase II effort is the critical next step. It will begin to answer many of the questions that remain, provide direction for future research and be crucial for future funding and programmatic decisions. In Phase II we will construct cable segments from carbon nanotube composites and test their general characteristics as well as their resistance to meteor and atomic oxygen damage. We will examine critical aspects of the Space Elevator design such as the anchor and power beaming systems, cable production, environmental impact, the budget and the major design trade-offs. Our previous work along with our Phase II results will then be introduced into the NASA mainstream effort through a conference and publication.'

7. PROGRESS IN THIS CENTURY

A core group of people seized the vision and started supporting the effort, striving to be first to get one built. However, Dr Edwards was regarded by most as the prime mover of the Space Elevator concept.

As is always the case with any development involving groups of people, conferences started to be held at intervals. Many conferences dedicated to the Space Elevator have been held, a recent one being the International Space Elevator Conference 2018, while the Third Space Elevator Conference in Washington DC in July 2004, where hundreds were in attendance, including Philip Ragan, co-author of the previous book with Edwards, is regarded as one of the most influential.

Dr. Edwards presents a session at the Space Elevator Conference in Washington DC. Photo by Philip Ragan.

At the same time that the third conference was held, the concept received a major publicity boost by being presented on the front page of Discover magazine.

"Memo to President Bush: scrap the Shuttle
and build a Space Elevator"

That was more than just topical. President Bush had announced an ambitious program to return to the Moon and expand America's presence in space, and yet only six months before, the Shuttle Columbia had disintegrated in re-entry to Earth.

It became increasingly obvious that a successor to the Shuttle program was needed.

Since then a number of people have contributed to the space elevator concept. Michael Laine established LiftPort with the intent of moving towards construction.

Dr Peter Swan has been an active participant in the International Space Elevator Consortium [ISEC] and the ISEC conferences. He has published a number of books on the space elevator, including Space Elevators: An Assessment of the Technological Feasibility and the Way Forward (2013) of which he was one of five editors.

The problem with efforts to date is they have been bottom-up approaches, in economic terms, rather than top-down approaches. If the President made the space elevator the priority of NASA and committed $10 billion to building one by 2030, with echoes of the Kennedy commitment to the Moon, then it could happen.

In the absence of this, two fundamental steps need completing before it will get serious funding: the discovery of how to produce CNT ribbons in commercial quantities, and the proof-of-concept usage of CNT ribbons on Earth, to build suspension bridges and the like.

In 2006 Edwards & Ragan published Leaving the Planet by Space Elevator, launched at the NASA X-Prize space hardware testing event at Las Cruces, New Mexico. It went on to be published in Japanese and Korean.

The Space Elevator looked an obvious candidate, but not just yet. The concept was too new for the conventional space-related agencies. So, for now we continue to use rockets but in something this radical it is possible that it may come about in a radical way, by a route not first considered. When will the carbon nanotube material get out of the lab and into commercial production? Back in 2004, we thought it was five years away. Here in 2018, sadly, it still seems five years away, but progress is being made.

By the time Stephen Baxter wrote his novel Sunstorm in 2005, the idea of a Space Elevator was becoming accepted as feasible. His concept was based on our 2006 book - having shared our work with Baxter prior to publication.

In his book, a near future human civilization has a Space Elevator operating from the ocean near Perth in Western Australia, an uncannily accurate representation of present plans.

What's changed in the concept since Sir Arthur wrote his seminal science fiction book, The Fountains of Paradise?

* Most importantly, we have carbon nanotubes now being produced in the laboratory to the required strength.
* We don't need a tower or building rising up from Earth to do this or an asteroid as the counterweight.
* The cable has been minimized so that it is now a light and flexible material.
* Many potential sites for the Earth terminal have been identified.
* The minimized cable is lighter than anyone ever expected. If it ever broke and fell, it wouldn't be a catastrophic collapse: it would resemble a newspaper floating to the ground through the air.
* The entire system has been laid out in terms of existing tech-

nology.

New technologies have a way of stealing up on their incumbents, and so it is with the Space Elevator. Most space programs, whether run by NASA, Russia, China, Japan, or a private venture such as SpaceX, assume that rockets are the only way to leave Earth. But things are changing: the Space Elevator has moved from science fiction to science possible.

What are we waiting for? The ability to produce CNT in woven ribbons, in commercial quantities, in runs of hundreds of thousands of kilometers. It has been produced in the lab, and in quantity in short length fibers, but not to the standard required, yet. Back in 2006, commercial production was forecast to be only five years away. Unfortunately, in 2020 it is still five years away, but we anticipate it will finally appear in the 2030s.

8. DISRUPTIVE TECHNOLOGIES

If it's such a good idea, why hasn't it happened already?

It takes time for disruptive technology of this magnitude to get implemented. It's not just the technology and the funding - we are still awaiting the desired breakthrough in CNT mass production.

It also takes time for people to get used to a new idea. It may seem a strange thing to say, as we live in a century where new ideas, new technology, keep appearing so fast, it's hard to keep up. But it's not only people in general; the people, scientists, engineers, who would work on such a project are already working elsewhere. In the case of rocket scientists, their work for the next thirty years could be mapped out already. It takes time for the space elevator to become the "next big thing" everyone wants to be part of. But its time will come, probably in a decade or so, in the late 2020s to 2030s.

The psychology of resistance to change is interesting. We only have to go back 100 to 150 years to find a time in history where change was slow, where horses had been the main method of traveling for thousands of years past. The Industrial Age was bringing in machines, mass production and electricity. Our ability to travel anywhere was brought on by three disruptive technologies we take for granted today: trains, planes and automobiles. What happened when they started out as ideas?

Given the expense and difficulty of using rockets, you would think that a new and better technology would be welcomed with open arms.

However, things are rarely that simple. In the past 60 years, a huge industry has developed around the production and servicing of rockets, a highly profitable industry. With a lead-time of 20+ years for the development of new designs, this industry is not keen to change direction. Many are reluctant to face the thought that the huge investment in rocket technology is only a passing phase.

NASA, too, works on 30 year plans, and is busy focusing on how to get to the Moon with rockets in the next decade, and a tight budget - again! That is, until a new President comes along and changes NASA priorities which happens like clockwork.

Understandably, it is too disruptive to say, "Just wait a decade or two, because the Space Elevator will then be capable of deployment and that will change everything". Their rejoinder would be: "but what if it doesn't work?" Anyway they have a budget and have to be seen to be spending money on projects now, so the momentum continues with work on existing technology.

It would have been just as logical for Queen Isabella, listening to the request by Columbus' for funding and three tall ships to find America, in 1492, to have responded by saying, just wait. In another five centuries we'll be able to fly there!

Alas, we humans do not make decisions as rationally as we would like to believe. Faced with major advances, we take time to get used to the idea; sometimes a generational change is needed. Not that this is always the case. Look how many technological changes we now live with:

* The internet
* Computers
* iPads and other tablets
* Smart phones
* Staying in touch with everyone 24/7

(A bit of humor here: in the Leaving The Planet book published back in 2006, we included the fax machine, and pagers, in the list of technology. How fast it changes!)

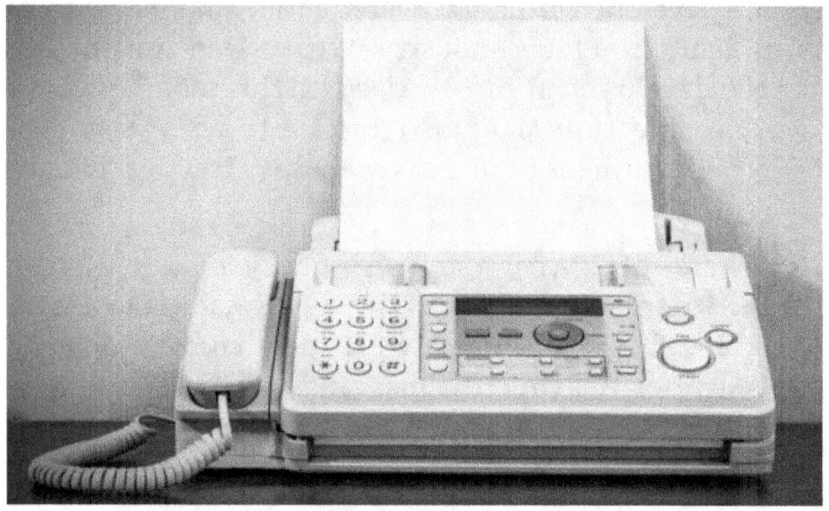

The Fax Machine: way back in history (the 80's)
this was the height of technology!

In whose footsteps are we following?

In their early years, advances such as trains, planes, and automobiles were viewed skeptically. We can go back into history to show the reaction they received, and yet how they replaced the existing paradigm.

"Plus ça change, plus c'est la même chose."

9. AIRPLANES

With cars and trains, 19th century people thought of themselves as technologically literate. But the advent of flight seemed, well, science fiction.

There were two sets of contenders for the first heavier-than-air flight: the Wright Brothers (USA) and Santos-Dumont (Brazil/France). Both were lampooned for their belief in heavier-than-air flight. A popular phrase of the last century was, "If God had meant for man to fly, feathers would be sprouting on his shoulders!"

The Wright Brothers first flew their airplane on December 17, 1903.

When they started, there was no such career as an airplane developer, in fact the Wright Brothers ran a bicycle repair shop and their first airplane evolved out of bicycle spare parts. Take yourself back to 1900 and imagine that you have money to invest in a project. A pair of bicycle mechanics asks you to invest in their new venture: they are going to use their experience in repairing bicycles to build a heavier-than-air machine that will fly! Would you give them the time of day?

It is worth remembering that some of our best advances have come, not from rich institutions, but from the "garden shed inventor" type, working alone on a crazy idea. All right, so many ideas are crazy, but those that are good have often turned out to be very good.

The first Wright Brothers airplane. Source
www.wright-brothers.org

Proclaiming "Wright Brothers Day" on December 13, 2001, the
President, George W. Bush, referred to the doubt that accompan-
ied their research: "Success had not come easily to the Wright
brothers. Many thought powered flight impossible; and skeptics
called the Wrights dreamers for even entertaining the idea. Un-
deterred, the Wrights spent years in study, analyzing the flight of
birds and experimenting with model airplanes in their bicycle
shop in Dayton, Ohio."

Some critics refused to attend flight demonstrations, on the basis
that they "knew" heavier-than-air flight to be impossible. They
still derided them as frauds for years after their flight! Alberto
Santos-Dumont didn't beat the Wright Brothers to it, but he was
just behind them in trying to fly an airplane.

No-one really believed that a heavier-than-air airplane was possible. Not that it stopped anyone from wanting to go up into the
sky. The preferred technology was balloons and Santos-Dumont
led the field in developing balloons for flight. In his view, people
would use mini-balloons to get around the city and drop in (literally) on their friends, and he did just this. For a decade it was a
familiar sight in Paris to see Santos-Dumont flying overhead in his
balloon.

He would hold dinner parties where the table and chairs were suspended aloft on wires, so that his guests could float in the air to get the feel of what it would be like to fly in the sky. This was entertaining for guests though he would no doubt have been considered eccentric. In the 1890's the general view was that flying machines would never exist.

He moved to the USA in 1902 and was credited with being the first to coin the phrase "airport". He called the process of flying "aerostation" using it as a verb, not as we would today use it as a noun.

What was the "establishment" reaction to the idea of flying? At the time, British Lord Kelvin was regarded as the leading scientist of the century. He arrived in the USA for a visit and lost no time throwing scorn on the ideas of Santos-Dumont.

Under the headline "Airship Is Useless, Says Lord Kelvin" he was quoted thus: "The airship of Santos-Dumont is a delusion and a snare. ... Why, an airship of that type to carry passengers – that is, passengers who would pay to be carried – would not be possible."

Santos-Dumont responded that it was necessary to demonstrate flight, and to get private enterprise in on the act, saying "That is the only way, I think, which will stimulate capitalists into building flying machines. That is the way the automobile was brought out. One was built and shown in Paris, and in a very short time it had been improved on and new ideas developed until we got what we have today, every other vehicle in the street is a horseless carriage."

On July 19, 1906, he took his heavier-than-air craft "No. 14" out onto the field and flew it for the first time without a supporting balloon. Dubbed "Bird of Prey" by the press, "every one present saw the wheels leave the ground and the machine travel forward, supported on air, six inches above the ground."

He trailed the Wright Brothers by 2 years, but the principle had been demonstrated: Lord Kelvin was wrong and a heavier-than-air machine could fly.

With hindsight it is easy to say that Santos-Dumont, and the Wright Brothers, were visionary heroes.

Returning to the speech by President Bush: "On December 17 of each year, we honor the Wright brothers for their contributions to our Nation. Their invention of powered flight made the world community more connected. We have since traveled to the Moon and back, and space shuttles orbit our planet. All of these amazing advances can be traced back to that windy day at Kitty Hawk when the indomitable will of the Wrights persevered to triumph over the supposed impossible."

It was once hard to believe that ordinary people would ever fly. Yet once the breakthrough had been achieved, look at the speed with which developments happened: the first Atlantic flight took place within five years, while ten years later, airplanes were flying in the First World War.

Now we routinely fly around the world in 24 hours or so. A century ago you would have been laughed at if you suggested this was possible.

10. TRAINS

Even before cars replaced horses, the railway system dramatically changed the world.

There is a parallel between the American transcontinental railway, and the Space Elevator. Until the building of the railway, travel across the USA was a difficult and costly affair. Building the very first railroad was enough to cut the cost dramatically. In terms of marginal utility, the first to be built is the most valuable.

There are other parallels between the two. A train line only has value when it connects two major points, a departure point and a destination. This is why the inter-continental railroad across the USA was constructed at a fast pace. The costs of construction continued to be accrued by the mile, whereas revenue and passenger numbers would not commence until the final destination could be serviced.

To defray costs, the railroads established townships en-route in an effort to encourage travel on the portion of the network existing. Before the inter-continental railroad was built, access to the West was difficult, expensive and risky. However, as soon as it was complete, the journey then cost only a fraction of what it did previously and this led to a dramatic increase in settlement in the West.

Early American Steam Engine. Source Smithsonian

We are used to traveling fast, at 100 kph or more on roads, at 1,000 kph in the air. Until the first train was built (Stephenson's Rocket) no human being had ever traveled faster than the speed of a galloping horse, 40 kph or so, and then only for short distances. It was commonly believed the human body was not made to travel faster and would fall apart! (The only way to travel faster was to jump off of a high cliff, but the result seemed to prove the point.)

Just as half a train line has no value, neither does half a Space Elevator. That too needs to connect two major points before revenue can be generated.

The parallels with the Moon and the rest of space are clear: before any meaningful settlement occurs, the cost of access has to drop dramatically.

11. AUTOMOBILES

The motor car may be taken for granted today, but in its early days, it was seen as a strange and expensive toy. It would never replace the horse and cart, would it? After all, horses had been used for millennia.

Owners of stables would have laughed if told that their stables would soon be obsolete! "Horses have been around as long as people", they would say, "and always will be! New-fangled cars are just toys for the rich." Throughout the 19th century the number of horses maintained a constant relationship to the number of persons in a city.

In the early days of the motorcar in London, the industry was stifled by the presence of the Locomotive Act of 1865. This law was introduced to require a walker bearing a red flag to precede motorcars, and car speeds in town were limited to 2 mph (3.2 kph). Commonly believed to have been an early safety measure, it was more likely a restraint-of-trade practice designed to prevent motorcars taking over from the horse and cart.

Early on, cars were seen as toys of the rich, as parodied in the classic book, Wind In The Willows, by Kenneth Grahame, in which Toad goes crazy for the latest invention, the motor car, and frightens horses and people alike as he drives dangerously along country lanes.

Toad of Toad Hall car

It took a generational change for the horse to exit stage left, but nevertheless it did happen. We now put it to you that rockets will go the way of the horse within a generation.

If we build it they will come.

It is ironic that people will resist the opening up of space travel, yet those same people will use their car, train and airplane in everyday life! We resist new changes, but we accept, as background, the changes wrought by the previous generation.

In the early days of powered flight, pioneers like the Wright Brothers and Santos-Dumont struggled to do the "impossible", that is, make an airplane that could fly. That was a century ago. Imagine being there while the Wright Brothers tinkered with their model planes. That is what it feels like right now to be involved with the Space Elevator.

As you read the story, you can share our excitement and wonder as the discoveries mount, and it dawns on us that we are making a change to the world every bit as significant as the airplane.

Can it be done within our lifetime?

Consider this: in 1957, no rocket had entered space, yet 12 years later men were walking on the moon! (Literally: men. NASA did not send any women to the Moon.)

Back then they had to invent the technology. This time around, we are mostly re-using existing technologies. Building the Space Elevator is a relatively easy task compared to the initial moon shot! All we need is that elusive, strong, CNT in commercial quantities, and we're off!

12. WHY DO IT?

Why do we ever do anything? Because we can; because we must. Modern humans rise to new challenges.

"The Earth is the cradle of the mind, but we cannot live forever in a cradle." - Tsiolkovsky.

Our "Pale Blue Dot" - Carl Sagan

As to why we should go into space in the first place, much has been written to make the case, such as in the writings of Carl Sagan and Robert Zubrin. Already convinced of the rationale, we anticipate that most of our readers are only too keen to see space exploration happening and are impatient at the seemingly slow pace of progress.

The reasons for expansion into space have four themes:

* Our innate curiosity and desire to learn new things.
* An insurance policy against our race on Earth being wiped out.

* The economic benefits of space-based trade.
* We'll never know until we go, i.e. the "because it's there" case.

So many writers and scientists have ably put the case that an attempt to list all of them would risk offending someone by inadvertently omitting them. However, we are motivated by the two mentioned above: the late Dr. Carl Sagan, and Robert Zubrin.

Dr. Carl Sagan played a leading role in the space program, and received numerous awards and medals for his achievements, and below we have quoted from his influential book Pale Blue Dot.
He did so much to promote the public understanding of space and science, and many people would remember him on television at the time that Mariner 1 first landed on Mars, and for his television series Cosmos. See www.carlsagan.com for more information.

Robert Zubrin is an astronautical engineer, ex-NASA team member, and writer of The Case for Mars in which he expressed his frustration with NASA and the slow progress towards human missions to Mars. Subsequently he set up the Mars Society that has pulled together like-minded persons worldwide in support of Mars manned missions. See www.marssociety.org for more information.

Why leave Earth to explore space?

We are driven by our innate curiosity and desire to learn new things, and we are already out there, exploring our solar system. We just need a cheaper and faster way of getting there.

Out of all the creatures on Earth, humans alone seem to possess the set of features needed to produce both curiosity and the means to satisfy it: intelligence, consciousness, and mind/motor coordination to test the environment around us.

In the 6,000 years or so that civilization has existed, human societies have exhibited bursts of discovery and inventiveness, starting with crops and farming, and great inventions like the wheel.

The Greco-Roman empire took inventiveness and philosophy to

new heights before the human race lost their achievements in the Dark Ages. Then, from the late 1600's onward, a new renaissance blossomed, spreading out from Europe, as new inventions like the telescope nurtured fresh curiosity, and the Industrial Revolution provided the means to produce ever more advanced discoveries.

With child-like pleasure, we delight in exercising our curiosity and learning new things, seeing new places. This driving force within us has created a civilization in the last century, the likes of which has never existed before in history.

When new frontiers have arisen, humans have always been only too keen to explore them. When the America's were discovered, a small group of pioneering seafarers braved many dangers to get there and establish colonies, the benefits of which eventually flowed back to Europe.

In A Case for Mars Zubrin argued that humans need to expand, to have a frontier. Zubrin also argues that we, as a society, are stagnating, that technological advancements are slowing, and that we need "frontiers" to encourage innovations and individualism. Zubrin argues this is a major reason for developing a Mars program. We would add that the argument applies equally to the promotion of all space exploration.

Why did we evolve? Why are we conscious, intelligent and curious? We can't answer that here, but it seems that our universe is structured to support beings with those characteristics. Given this, it is inevitable that we will want to expand into space and explore other worlds. All we need is the means to get there.

An insurance policy

It is amazing that we are here, on our planet, alive and functioning. We face so many dangers, some admittedly of our own making.

Sadly, when it comes to climate change, global warming, pollution and the destruction of habitats, men seem quite happy to destroy the planet as a trade-off for their own personal profit. The opinion is sometimes voiced: the money spent on space exploration would be better spent on looking after our planet. But it's not an either-or choice, though if you want one, how about we stop spending on wars and weapons and use that money to improve the planet?

Looking after our planet is a whole other issue, perhaps the most important issue of this century, and we strongly support that. If we ever reach a time when we have to move to Mars simply because we've ruined Earth, it would be a sad day indeed, but this isn't even a practical reason to explore space, for even an Earth with a ruined environment would still offer better opportunities for life, than cold and distant Mars.

On Earth, we are now aware how fragile our environment can be, at risk from dangers like pollution, ozone depletion and global warming. Our population numbers and profligate consumption pose dangers to the mutual support systems of Earth. We seem willing to eat or push other life forms into extinction. We possess the means (i.e. nuclear weapons, wars) to destroy the planet. We desperately need to solve these problems. Some of them can be answered by science, while some, such as our strange propensity to fight and kill each other, call for changes in human behavior that seem as yet unsolvable.

From space, our planet is always at risk of being hit by large asteroids that can ruin our environment, as the dinosaurs discovered. Asteroid strikes fall into that category of risk that combines rare occurrence with enormous losses. The next strike might be next year or a million years away, but when it happens, it can be the end of life on the planet. Who knows what other dangers we face? If we have colonies on the Moon and Mars, at least this is some insurance for our continuation if an asteroid strikes.

However, even if we solve all problems on Earth, those dangers from space are ever present. As long as we all live on Earth we have "all our eggs in one basket" to quote the old proverb.

Carl Sagan, in his book Pale Blue Dot, recognized how fragile and small our home is, by comparison with the universe and that, sooner or later, our lives on this planet will end, and our civilization with it unless we have spread our wings across the galaxy. The book title came from a photograph that he commissioned from the satellite Voyager 1. By February 1990, the satellite was on the way out of the solar system. It was instructed to turn the camera back towards the Sun and Earth, and take one last picture. Speeding away at 60,000 kph it captured the classic photograph that appears on page 2 of his book.

From the edge of our solar system, our planet can be seen only as a dot in the sky, a pale blue dot at that, the color blue being the only clue that life even exists on that dot. Everything we are, our civilizations great and small, our families, our passions, our hopes, are all contained within one pale blue dot, an insignificant dot of billions of such dots in the universe. If this dot was wiped out, it would matter to us, but its' removal would be barely even noticeable elsewhere.

As Carl Sagan put it: "Since, in the long run, every planetary society will be endangered by impacts from space, every surviving civilization is obliged to become space faring – not because of exploratory or romantic zeal, but for the most practical reason imaginable: staying alive. ... If our long term survival is at stake, we have a basic responsibility to our species to venture to other worlds."

It's not only an insurance policy against asteroids or volcanos. In the far distant future, our planet will cease to exist. Life, in this universe, only has a future if it spreads far and wide into the universe.

The economic benefits of space-based trade

We are all familiar with the economic mantra of growth, and the amazing growth in most global economies over the past 70 years has taken most of the world population from subsistence living, to comfortable living.

In the sixties, the outlook and projections for overpopulated countries like India and China were dire. By today, famine, disease and disaster were forecast to be the lot of these countries. Yet, in reality, fewer people are in poverty in those countries today, compared with the sixties. Having tasted the benefits of economic growth, those people are in no mood to give it up.

As we start to reach limits on the level of growth that can be squeezed out of one planet, we are faced with the obvious: growth will, at some point, only continue if we can find other raw materials and resources, and that means expansion to other planets and moons.

Once we have the means to travel into space at an economic cost, we have the trifecta to justify it on an economic basis. They are:

* Low cost access to space travel.
* Surplus population willing to move into space.
* Access to space-based resources.

We are not talking about wholesale movement of millions of people into space, though there is no reason why this could not occur. The establishment of the Americas was achieved by the movement of relatively few people who took to the seas in galleons. Most of the European population actually stayed in Europe. A relative handful of people established the new colonies. Growth was achieved organically with new families and births in America, and by the compound growth of a steady stream of migrants each year. That was enough to produce, over three centuries, the USA that is now the largest economy in the world, and its' powerhouse.

The same outcome can be expected from space travel. With low cost access to space, it will not be long before the seed capital is generating a return. If we are right, the return will be incredibly large within only a few decades.

Countless science fiction writers have painted the future, postulating space based empires stretching over planets and nearby stars within a few centuries, and eventually human civilizations and trade stretching across the galaxy and beyond. It only took three centuries to create the USA; we figure that three more centuries will see humans across all of our solar system at least, and pushing out the frontier to other star systems.

Does economics give us a reason?

Can an investment in a Space Elevator really make an economic return? A "yes" answer must be part of the reason to do it.

Consider the transcontinental railroad and how that benefited the United States. The entire western seaboard is now a major economic force because of the railroad. Yet, communications between east and west were negligible until the first railroad line was built. For almost a century the railroads carried goods and people and earned profits for their owners.

Consider how the current rocket-launched satellites have enabled global telecommunications and Earth climate-monitoring systems. These items have produced dramatic progress and benefit to our society over the last 40 years. Even at rocket-based prices, launching these satellites has been profitable.

The Space Elevator technology is still a work-in-progress, but it is maturing, soon to be available. We have researched the system sufficiently to have a good estimate of the cost, at least as a broad cost range.

How much of a difference does low-cost access to space make?

Affordable space travel is great, but not if you have to send the equivalent weight of 142 trucks just to get one person up there. However, there is good news. First, we can lose the burden of transporting massive amounts of fuel. Most of the fuel cost of rockets is expended in breaking through the lower atmosphere, and our Space Elevator, though much slower, can do the same trip with a fraction of the fuel. On top of that we don't need the massive rocket either, so we eliminate the old constraints that the rocket presents.

No longer do space trips have to be large, occasional or expensive. We are talking a number of elevator car departures per day, every day, giving us a vast capacity to put people and material into space on a scale never before contemplated.

Cheap, easy access to space opens possibilities. Just as air travel makes a dramatic difference to the world when the cost of a round-the-world ticket falls over a generation from $30,000 to $2,000 so too the attractiveness of space depends on the cost of getting there.

It is a simple cost-benefit analysis. The Elevator is so much cheaper, and the environmental impact so much more benign, that it really is the only way to go. There is no better way on the horizon!

What happens if we don't build it? Rocket manufacturers have a vested interest in keeping rockets as the state of the art technology, but that means keeping the cost of space access high. It would remain an exotic activity until such time as another country or corporation takes the reins.

Return on investment

By now you will appreciate that the Space Elevator is a tremendous undertaking. It is on the same scale as the Apollo program, the Panama Canal or Hoover Dam, though even these pale into insignificance in comparison to the achievements of the Chinese economy over the past 30 years.

Return on investment then depends on the income to be earned from space travel and ventures. There are significant returns from cheap access to space. Here we can touch on:

* Telecommunications satellites
* Astronomy

* Expanding our resources
* Colonization of the Moon and planets
* Solar power transmission
* Protection from asteroids
* Other spin-offs
* Jobs and Space Tourism

Before we start on new challenges, reflect on the things that are profitable already:

Telecommunications satellites

Even with the enormous present day cost of rocket launches, it is still profitable to put satellites into space for telecommunications, Earth surveys and Earth imagery. At the lower cost base provided by the elevator, launching and servicing of satellites will become easier and cheaper.

Astronomy

As a country, we also accept the cost in relation to less tangible things like exploration satellites and telescopes. Space-based telescopes have wide community support and win much praise for the impressive images of the heavens. The ability to support Moon-based observatories, particularly on the "dark side" of the Moon, will be popular with astronomers.

If we can offer those same services at lower cost, then we would expect demand for those services to expand. Therefore, we have two viable markets that are already in existence.

Expanding our resources

The number of people and objects that can be put into space will increase dramatically when the first and subsequent Elevators are deployed. Additional revenue sources will flow from this, and they can be grouped as:

* Charges for lifting materials into space, and return.
* Mining of minerals on the Moon and other objects.
* Charges for travel by persons into space, and return.

* Transmission of power back to Earth.
* Critical space missions such as diverting asteroids.

Some of these benefit Earth directly, as in beaming power back to Earth, or diverting dangerous asteroids, using the lifting capacity of the Elevator to put defensive missions into space.

Other benefits arise from exploring new frontiers, creating new colonies and wealth in space, to boldly go where no human has gone before.

We'll never know until we go

The exploration of the new continents of the Americas was undertaken, firstly, to find a new trade route to the East Indies. Then, products such as gold, silver and sugar were shipped back to Europe. Did Columbus set off on his voyage thinking that it would lead to space flight? He would not have foreseen any such thing.

When we embark on new ventures, it is only after the event that we will know what we achieved. While we can write up all of the reasons that are apparent today, historians of the 22nd Century will be able to write about what really happened, and what great new discoveries or events flowed from our actions. As with all past explorers, we continue to have faith that good things will flow from our efforts.

Our pale blue dot, our Earth, is but the tiniest home in the galaxy. Carl Sagan couches the argument for space exploration in terms of survival, though we personally are just as happy with "exploratory zeal", or exploring "because it's there". If you are the type of person that thinks "awesome" at the prospect of mountain climbing, abseiling, rafting, etc., then you know what we mean.

Lack of progress towards a space-faring civilization has been a source of frustration to all those who want to see it happen. Given the prohibitive cost of conventional rockets, our chances for realizing space travel on a massive scale are poor, if this is all we have. Space looks like a luxury to the general population and it is always difficult for them to watch their hard-earned money going to taxes that fund such a luxury.

There must be a stronger motivation, something that happens on timescales that we humans function in. What can we tap into that will get humanity excited about space travel and lowering its' cost? There are drivers like national pride, economic return, and fame. These have worked in everything from daily lives to global endeavors. If only we could reduce the cost of space travel, then these factors could come into play.

Fear of coming second to the Russians inspired the drive to the Moon in the 60's. If America wakes up one day to find that another country is deploying an Elevator, the fear of losing control of space will provoke the same reaction and start another space race. Meanwhile, as each year goes by, our 'pale blue dot' becomes just that little bit smaller for everybody.

13. WHY A SPACE ELEVATOR?

We are already in space. We get there with rockets. The private sector is getting into the act, with tourist trips into space in the offing. So why do it with a Space Elevator?

Our current rocket launch technology depends on a handful of launch sites. It is not widely understood that our hold on space travel is tenuous. The much-heralded ISS depends on servicing from private and Soyuz craft, since the retirement of the Space Shuttle.

Hurricanes are a major threat to the Kennedy Space Center at Cape Canaveral. At the least, storms cause a shut-down of operations, though so far, hurricanes have not scored a direct hit and damaged the Center. Hurricane Matthew passed within 26 km of the Center in 2016. Reports indicated there was limited roof damage and debris around the facility, but all of the center's equipment — such as launch pads and its assembly building — were constructed to withstand wind speeds of 200 kph. The wind around the Kennedy Space Center hit an average of 135 kph, so the facility managed to escape the feared level of damage, this time.

Will private enterprise lead the way into space? A number of companies are already in the business, with launches to the ISS or at least, clearing the atmosphere. This is being done by aerospace companies like Blue Origin and Virgin Galactic. However, there is a gap between such sub-orbital planes, and traveling to the Moon. The planned tourist flights are only capable of a brief sub-orbital flight at a height of 100 km or so above Earth. They only spend a few minutes at that height, which is where space officially "starts".

Why only a few minutes? That word sub-orbital is the clue. It is not going fast enough to stay in orbit. They cannot orbit the Earth, even at that low height. Like a firecracker, they goes up, hover in space for a moment, and then fall down again, flying back to Earth.

There is a big gap between achieving a sub-orbital flight, and going into orbit, or even visiting the ISS which is 386 km up. It is the same equation that bugs all rockets. The extra fuel required to reach orbit would call for a vastly larger craft. In turn, the weight of the larger craft requires more fuel still, and so on. It isn't just a matter of making the rockets a bit bigger. Achieving orbit with a rocket calls for an order of magnitude increase in scale. Going to the Moon, an order of magnitude bigger again.

While it is possible that, based on the design experience of such craft, future rocket design might result in rockets that bring the cost of space launches down, we are talking of reductions of only a few percent off present day costs.

We currently only have a multi-billion dollar rocket-based program that is an expensive high-risk technology. We need something better if we are to stay in space. To the best of our knowledge today, we can only envisage two ways of getting into space, by rockets, or by climbing up an endless ladder. So, that is where our Space Elevator comes in. It is the only viable alternative.

Colonization of the Moon and planets

As little as ten years ago, the prospect of colonizing the Moon and Mars was still a generation away.

With our new Space Elevator, however, we can now do it for real! The arguments continue for bypassing the Moon and just going to Mars, but with our cost structure, we can happily fund colonies on the Moon and Mars.

Colonies on the Moon

Each Space Elevator will be able to deliver 1500 tons of material to the Moon or Mars every year. The mass of a lunar base with a population of roughly 450 people is around 4,000 tons with an annual resupply rate of about 1,800 tons. So, using part of the capacity of the first few Elevators it would be practical to have a lunar base with a thousand people up and operating for much less than the current NASA budget.

At this level we could have a full infrastructure on the Moon, a self-sustaining society, producing products and resources for use back on Earth. People would be born and live their lives there. Now what happens when we build this infrastructure and have valuable activities occurring like mining, power production, launch facilities, astronomy, Earth observations, and manufacturing? Private enterprise sets up shop and expands the population and infrastructure. Private enterprise has much more financial power than the government in this situation and will quickly drive development. The Space Elevator effectively collects a fee on all this activity, since it is the conduit by which all materials and personnel must travel, just as airports and airlines charge a fee for travel.

Compare this to the European ocean-going ships that crossed the Atlantic to North America or the transcontinental railroad. Both end points grew and profited. Boston competed with New York, Liverpool competed with London, to the benefit of all.

If we speculate on what will happen, we have a lot of confidence in stating that after the first several Elevators are built expansion will be rapid and within two decades we will have substantial cities on the Moon and Mars with thousands of people. Imagine that - thousands of people living, going to school, socializing, and producing products and services. A new "frontier" is a reality, the first new colony in hundreds of years.

Colonization will start with the Moon, we confidently predict, but it will quickly spread once the techniques and value are demonstrated. Mars and the asteroids will be colonized next followed by the moons of the larger planets.

Edwards' grandmother once stated that in her life the world had gone from covered wagons to men on the Moon. That was a time span of 86 years. Our children will say they saw human society expand from a single world to dozens, from a time when a few people a year might venture into space to one where the corporate IT technician might get assigned to a job on Mars. In another 86 years, there will be dozens of major cities on the Moon and Mars, a large number of asteroids will be under development and there could be human outposts floating in the atmosphere of Venus and on several of the moons of Jupiter.

The first stepping-stone to this impressive future will be the establishment of a lunar base, and with the Space Elevator this is totally feasible for less than we are currently paying to NASA to maintain our existing rocket-based space technology.

Earth services

We will eventually run short of oil but energy is all around us. We are starting to use solar arrays here on Earth for power but things like night, clouds, and the atmosphere limit their efficiency.
Well, in space there is no night, no clouds and no atmosphere. Solar arrays work much better in space. Now think about putting square kilometers of solar arrays in space to collect the limitless power of the sun and sending this power down to Earth on a laser beam or microwave. No pollution and it won't run out. The electricity generated can run cars, factories, homes, everything.

Great idea, so why aren't we doing it? We would if we could; people have been working on the idea for 30 years. In 2018 came news of a Chinese plan to light up cities from space, starting with Chengdu.

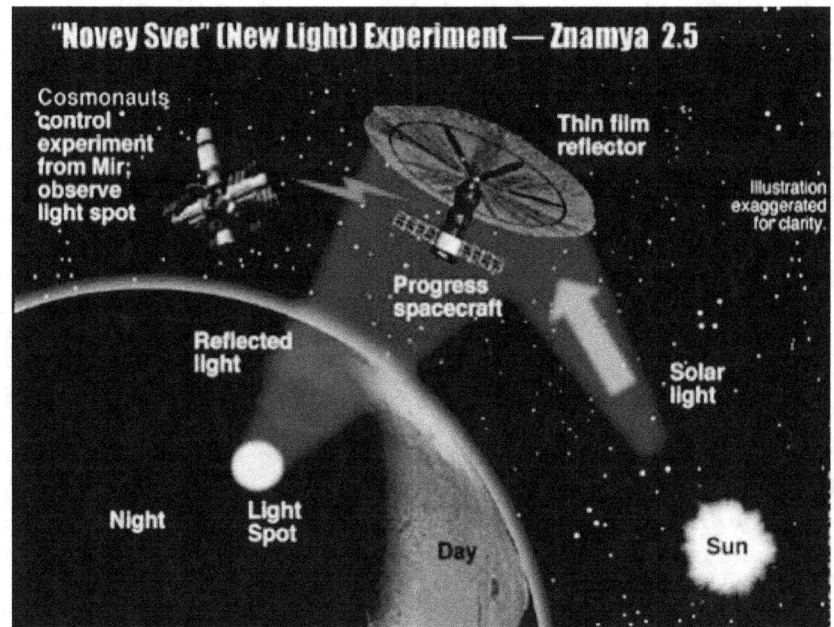

Chengdu New Light

Reports phys.org: 'China is planning to launch its own 'artificial moon' by 2020 to replace street-lamps and lower electricity costs in urban areas, state media reported. Chengdu, a city in southwestern Sichuan province, is developing "illumination satellites" which will shine in tandem with the real moon, but are eight times brighter, according to China Daily. The first man-made moon will launch from Xichang Satellite Launch Center in Sichuan, with three more to follow in 2022 if the first test goes well, said Wu Chunfeng, head of Tian Fu New Area Science Society, the organization responsible for the project. Though the first launch will be experimental, the 2022 satellites "will be the real deal with great civic and commercial potential," he said in an interview with China Daily.'

The problem has been that the initial launch costs are still high for putting the solar arrays in space. If the launch costs were lower, then initiatives like this would be quickly copied around the world, as well as for solar panels. The Space Elevator will lower the cost to an economic level. So we dedicate a few Elevators to building solar power satellites and put up these huge arrays. The dependence on oil disappears. Pollution dwindles. Our children and their children will have a consistent source of energy for all their needs, and it benefits the environment too.

Protection from asteroids

Have you seen the movies Armageddon or Deep Impact? In spite of the artistic license that these movies adopt, they highlight a critical issue. There are asteroids that could impact Earth causing global destruction. The likelihood of an impact is believed to be low in the short-term but it is clear what will happen if a large object does hit Earth. These large objects are extremely difficult to deflect and to do so will require extensive assets in space. If an asteroid were headed for Earth right now we would have little chance of protecting ourselves.

With extensive spaceship systems at geosynchronous orbit and the Elevator to support them, a system could be put in place to meet a dangerous object early, when the minimum energy is required to deflect it. Even in this case, to deflect an object a kilometer across could require thousands of tons of fuel and hardware. This would be challenging for the Elevator but much more viable than with any current system in place today.

Other spin-offs

Construction of the Elevator has immediate tangible returns as discussed above but these are only the benefits we can readily see. There will be other terrestrial spin-offs, as happens when any new technology is developed. The carbon nanotube cables will have a strength x180 that of steel by weight. They will likely replace all other materials where tensile strength is required, e.g. suspension cable bridges. Many applications for carbon nanotube composites exist today.

Jobs and tourism

As a society we can look forward to the creation of a new class of employment as space related activities increase. It is not only about the people who go into space: many more jobs will be created on Earth, supporting the Elevator, creating products to go into space and supporting the ventures and colonies up there. The growth of the IT economy in this century creates new wealth and so too will the growth in space industries.

Jobs in space

Tourism will, in itself, be a major revenue earner for the Elevator. Consider how tourism benefits low-cost airlines on Earth. Apply that model to space tourism.

We, as a society, may be approaching some limits. Our planet may not be able to support our growing population. Our fuel supplies may get depleted. At the very least, rising fuel prices make the usage of rockets even more expensive. In a world of rising resource prices, the relative economics of the Space Elevator will keep improving.

We can't guarantee that a Space Elevator will solve all these issues, but it sure would be nice to have these space-based options when trying to solve them. Until we get out there, we won't know what other solutions a space-faring culture could provide.

Enough of the serious stuff: What exciting new prospects are in store?

Maybe we will see the low-gravity football league in 2100, zero-gravity football, the rocket regatta of 2090, the exclusive Gates Resort on Olympus Mons, the ring-world community completed in 2120 complete with Zero-G Ocean and continuous daylight, or the first generation ship headed for the planet at Proxima Centaurus.

Sound fanciful? If history teaches us anything it is that we cannot predict the full magnitude of what can and will be achieved.

The question we placed at the beginning of this chapter was why we should build the Space Elevator. When it does become reality, it will move and advance our society as few other developments have. The Space Elevator may literally save our children's lives and benefit our civilization is vital ways.

14. A PARADIGM SHIFT

The reasons, then, to use a Space Elevator to go into space, are that it changes the rules dramatically and reduces the cost to an economic level.

The Space Elevator is a disruptive technology. Disruptive technologies are not gradual change. They rewrite the ground rules. If you are playing the old game of "rockets" you are out of it when an elevator offers launch capacity with a "98% off" sale tag.

Disruptive technologies are the key to changes in nation strengths. Steam engines gave Britain its power in the Industrial Revolution and turned it from an average European country into a world power for a time. Cars and airplanes were the disruptive technologies at the start of the 20th Century. Smart phones mean newly developing countries can leapfrog landline phone systems.

The first country or entity to successfully implement the technology will quickly gain economic and political power in the form of controlling the primary access to space: telecommunications, energy, the moon, Mars, asteroids and a myriad of new markets. This will drive the history of the 21st century.

The cost of inaction

Many of the comments in this book are predicated on the USA building the first elevator. But what if it isn't the case?

Looking through history we find that there have been endeavors that were much more extreme for their time. Consider the Pyramids or the Great Wall of China. In each case there were overwhelming driving forces to undertake it. The pyramids were constructed as a part of the social and religious structure of the ancient Egyptian society. It was part of who they were. The economic returns and long-term impact were probably not a consideration.

The Great Wall of China was a piece of military infrastructure built to protect the Chinese from their unwanted neighbors, the Mongols. The wall was built to save lives and their way of life. For every successful "Great Wall" there have been other projects which, with hindsight, should have been completed but weren't, with devastating effects.

One classic example comes from ancient China, when they failed to exploit a new frontier. In the early 15th century it [China] sent treasure fleets, each consisting of hundreds of ships up to 400 feet long and with total crews of up to 28,000 across the Indian Ocean as far as the east coast of Africa, decades before Columbus's three puny ships crossed the narrow Atlantic Ocean to the Americas' east coast. Why didn't Chinese ships proceed around Africa's southern cape westward and colonize Europe, before Vasco da Gama's own three little ships rounded the Cape of Good Hope eastward and launched Europe's colonization of East Asia? Why didn't Chinese ships cross the Pacific to colonize the Americas' west coast? Why, in brief, did China lose its technological lead to the formerly so backward Europe?

The end of China's treasure fleets gives us a clue. Seven of those fleets sailed from China between 1405 and 1433. They were then suspended as a result of a typical aberration of local politics that could happen anywhere in the world: a power struggle between two factions at the Chinese court (the eunuchs and their opponents). The former faction had been identified with sending and captaining the fleets. Hence when the latter faction gained the upper hand in a power struggle, it stopped sending fleets, eventually dismantled the shipyards, and forbade oceangoing shipping.

One reason this book is written in English and not Mandarin Chinese is that political misstep almost 600 years ago. Consider that in our current environment we, as a society, seldom look beyond a few years in the future. The economic and political power of China is growing and likely to overtake the USA during this century. It is unlikely China would make the same mistake again.

We are not seeking to create another version of the arms race, or inter-nation rivalry. It would be nice to learn from our mistakes and, this time, advance into space in a cooperative spirit instead. However, it seems an unfortunate fact of our modern life that fear is a big driver of national advancement.

Bear in mind that the entity that builds the elevators will be the one that controls the supply of energy from space, and the revenue from space – a real good reason to be the one who builds the elevator. We can see the level of influence and attention a few small countries in the Middle East have garnered due to their energy resources as compared to similar countries in Africa or South America.

Here we will boldly claim that we are at a critical juncture in time. In the next twenty years, a paradigm shift will occur. An entity will grab the reigns and step on the path to become the dominant global leader for the coming century. The United States could continue in the dominant role but the reigns could just as easily be picked up by Europe, Japan, Russia, China, and Australia or even by a corporate entity.

What we are confident of is that the next few years will determine how the history of the next century will be written. It will happen!

Our civilization is already determined to expand into space. As Carl Sagan said, our planet is but a pale blue dot in the universe, and if we just stay here, we are certain, eventually, to follow the dinosaurs.

Once the cost of accessing space is lower than the returns earned by traveling there, we will open up an entirely new market. Private enterprise will ensure that this market opportunity is seized, just as the railroads opened up the west of America.

It is unlikely that anyone argued the case to build the transcontinental railroad so that California could become the fifth largest economy in the world. Likewise, Sputnik was not launched by the Russians to promote 500 cable channels, satellite weather maps or a billion-dollar science fiction industry. So too with the Elevator: what unimaginable benefits will accrue to the Earth and to our grandchildren?

Surely the best reason of all to build a Space Elevator is that you and I stand a realistic chance of experiencing space travel after all!

Carl Sagan wrote:
"The notion of our descendants living and working on other worlds, and even moving some of them around for their convenience, seems the most extravagant science fiction. Be realistic, a voice in my head counsels. But this is realistic. We're on the cusp of the technology, near the midpoint between impossible and routine."

There is an exciting future in the sky, just waiting for us!

15. IMAGINE A TRIP ON THE SPACE ELEVATOR

Enough of the technicalities. What will it be like to travel into space on the elevator? Let's imagine the experience of being a passenger on the elevator, traveling into space.

Ever since Concorde stopped flying, the best air experience has been First Class on a jet. The lucky (and rich) few have had a ride to the ISS, and advanced planes may soon give a space-grazing experience: a brief ascent 100 km or more to sample the edge of space.

But with the Elevator in place you have the chance to be one of the first tourist passengers to ride the Space Elevator. Now that will be a real conversation starter at a party!

"You've been on holiday in Tibet? Oh yes, I saw that from space when I went to the Moon." Try and top that one!

What will it be like?

The first difference, of course, will be cost. It will still be expensive, but expense is relative. A ticket may cost from $20,000 (dinner at LEO) to $1,000,000 (lunar vacation) but that is much cheaper than a $40 million trip to the ISS!

Secondly, the trip will be less risky and more tranquil by comparison with rocketry. The levels of safety will be closer to those of a jet plane than a rocket, so you can be more relaxed about the trip.

So let's follow the trip through the imaginary blog of Livi, Olivia, who is one of the lucky few to get an early ride to the Moon as a tourist. The blog starts on April 10, 2033. Livi is soon to be 21 and this trip is an expensive 21st birthday treat from her parents, Matt and Annie. Matt is Manager at the company which built the first elevator, so Livi is a lucky girl indeed.

(Why Livi? A heads-up here: this is an extract from my scifi series "Enough Rope", a story of how it happened in the next fifty years, incorporating the space elevator, and Livi is one of the characters in those sci-fi books. So, this is a preview: her story.)

16. MY JOURNEY TO THE MOON: OLIVIA'S BLOG

The Departure Gate

Hi I'm Livi and welcome to my blog/vlog, a very special blog/vlog about me going to the Moon on the Space Elevator! So, yeah, I'm soon to be 21, as you can see from my bio info on your screen and you can follow me, like, on my vlog, YouTube, Facebook Live, Instagram, Twitter, you're already following me, anyway, right? I know you can see me, but they want me to write a record of the trip for posterity too.

hashtag #LiviGoesToTheMoon

April 10, 2033 0645 hrs

So it's, like, April 10, 2033, early in the morning. All the way on this trip I'll be reporting in Hawaii standard time, which is the time kept on the space elevator and at the colony on the Moon.

I'm on a cruise ship heading for the place where the space elevator takes off, the Earth Space Port. What will it be like? In many ways, like flying first class on a plane, with a big difference, well, two actually, I'll be in space, really, pitch black with stars outside, and as we travel, I lose weight, yeah, then get to experience weightlessness.

Am I scared? Absolutely, for sure. But I'll get over it. First time

I flew in an airplane I was so scared, that feeling of, if it crashes I'm not in control. But after a while you don't even think of your plane crashing, it's just routine. Anyway, on and off, over the past year, I've been getting training, doing simulations and stuff. So much more you have to know, like how to get a spacesuit on, in a hurry, and, like, you REALLY pay attention to the safety demonstration lol. In an emergency, you can lose air so you have seconds to get into the safety compartment and you really want to know where that exit door is. Plus, in an emergency you have to be able to take over the operation of the cable car, not that it's ever likely, but in space, well, you have to be prepared for anything. Just imagine if you couldn't board an airplane without having done a course on piloting the plane - that sort of thing.

OK, having written all that, if I'm scared then, even more, I'm excited! There's over 50 people up there already, on the ribbon, on the Moon, but I'll be the first certified tourist and my job is to sell the idea, show you what it's really like. Someone out there is running threads about nepotism and how it's unfair I get the role (I'm looking at you Ginny) but I'm doing it, so get over it.

I'm outside, on the deck of the ship, now. We're heading for a pit stop at Honolulu, where there's a crew changeover of Elevator people, then we carry on south, to the floating Space Port, near Starbuck Island, and I hope that means good coffee lol. After the crowds of LA, being on the open ocean is amazing. Early, sun rising in the east, just us, open ocean (if you don't count the drones hovering, filming me right now - see me wave?) and blue sky above us. Amazing to think, tomorrow I'll be flying through that blue sky into space.

Do I get points on a loyalty program for this lol?

April 11, 2033 1340 hrs

Hey. We're somewhere near Starbuck Island, close to the Equator, hot weather, clear blue sky! I've been getting a swim in the pool on deck, last one for a long time! Not so we land on the island, the

Space Port is floating out at sea.

I can see the Space Port now and it's like, wow! It's a collection of floating platforms, looking like a huge oil rig, and there's two cruise ships docked alongside one platform which must be the hotel and departure terminal. Add in container ships, cranes and stores, it looks for all the world like a port somewhere. Well, it is, I guess, but in the middle of nowhere. We're about 2 km away still. Funny thing is, I can't see the space elevator ribbon. It's so small and thin but I thought I'd be able to spot something by now.

What I can see - excited - a cable car coming down. It must be on the ribbon but from here it looks like a big balloon sort of hovering in space, it's going so slow! Ah, someone wants me. Time to pack and disembark, apparently.

This ribbon is owned by the USA and run by the military, so security is tight. Something struck me as a bit strange: before I can board the space elevator, I go through customs and get an exit stamp on my American passport!

Apparently, the GeoStation and the Moon are treated as external territories of the US for travel purposes and I need a passport, would you believe. I have visions of finding a duty free store when I arrive at the Moon, lol.

We wear a space suit to start this journey, though later tonight, we can take it off.

However, the "safety demonstration" is much more than the brief demonstration on an aircraft.
Accidents might happen, and the crew is trained in ensuring the safety of the car, including attending to such things as fires onboard. Plus, everyone, even me, has had the same training as the crew so in theory, we can also deal with emergencies.

The major risk is penetration of the outer skin of the car by a meteor or piece of space junk. Most such items are only as big as a grain of sand, but the creation of even a small hole in the skin could risk depressurization of the internal cabin.

Hopefully that event won't happen on our trip, but if it does, it's not just a case of having an oxygen mask dropping from the ceiling. If pressure drops I would need to don my full space suit, quickly. That procedure has been rehearsed in full before, and we'll rehearse it again several times on the journey out, so that we are comfortable with how to do it in different levels of gravity.

April 12, 2033 1100 hrs

Hey all. Here I am in my seat, inside the cable car pod, wearing my space suit! There's a countdown and soon we're off. I got to touch the ribbon last night. It's so weird, like meters across but nothing wide! So black, like old fashioned carbon paper, almost a pun, when you think, hey, this is literally carbon paper! It looks like a supersized hair ribbon stretching into the sky, but it just disappears in the air. Except at night, when it is lit up, like a string of Christmas lights above your head.

Sitting here is sort of like first class on an airplane, with a big comfy seat, though a very light frame. It folds flat for sleeping on later. Inside, the layout is like a three storey house, complete with stairs and bathroom …. wish I hadn't thought of that. We are in a pod, like a golf ball fifty meters across, so it doesn't feel as cramped as a jet. There are three pods, a strong central one, containing all the engines, wheels and stuff, plus the windows, exit doors plus the safety compartment, then an inflatable pod on each side. We are in one of the inflatables, while the other contains goods, luggage, fuel, air, etc.

Can't see out from here; once we are underway I can go to the window in the central pod and look up or down. JOLT! We are underway. It doesn't feel like a rocket or an airplane, though, much too slow. More like being on a train slowly pulling out of Grand Central.

Now you can feel the ribbon is flexible. My stomach is complaining as we move about sideways, as well as traveling up, a bit like how I feel on a yacht. There is a large screen showing the view downwards - the Earth Port is shrinking as we leave it. One thing is just like an airplane though - the Fasten Your Seat Belt sign lol, and just like on a plane, I'm waiting for it to be extinguished, so I can visit the bathroom haha.

The days of high-g acceleration and accelerometers passed away with rockets. That is good for health reasons: the human frame does not take kindly to 6-g acceleration. Instead, it feels more like driving down the freeway and indeed, after a while, the sense of "going up" will change to a sense of "going sideways", they tell me, as the car drives along the Elevator ribbon looking almost like a train on a monorail at Disney World.

April 12, 2033 1630 hrs

Hey again. It's about five hours in and we are in space! About 1,000 km altitude, already we've passed the altitude the old ISS used to orbit at, so I'm the 92nd human to travel into space, above the LEO orbit of the ISS, so they tell me.

We've stopped here, adjacent to a cable car which is heading down. Above this altitude we need radiation shielding. What happens is this: the car heading down has its inflatable pods surrounded with two meters of water at the perimeter, used for radiation shielding. It doesn't need this lower down, and it saves weight if the water gets pumped into our cable car pods, which also means we haven't had to haul all that water weight up here. So the AI computer has docked the two cars to carry out the transfer.

We've been give the ok to get out of our space suits and into comfortables. The space suit design has improved on the olden days; these are designed so you can get in and out of them quickly, real quickly, kind of like a well designed ski suit. But wherever I go, I have to take my folded up suit with me, just in case. Space, so they keep telling us, is not forgiving.

Meanwhile, I'm in the safety compartment where I can see out the windows. I can see people in the downdraft car waving back at me. But, hey wow, you should see the up and down views out the window! Well, you can if you are following me live, or on video.

Looking down, the Earth still looks close, but, compared to the view out of an airplane at 10 km up, definitely, we are further away, enough to see the dark of space. The Earth Port is just a tiny speck down there, just visible if you look in the right spot, while the ribbon itself might as well be transparent.

Back to the side window, I've got the classic view of the curvature of the Earth, the bit in shadow in the distance, coming up as night approaches. It's not like the old ISS videos, where the Earth rotates by. The ISS was in orbit, whereas we are stationary relative to the Earth. I can also see our huge solar panels, opening up and being deployed, while the water transfer continues.

Looking up, it's a jet black sky, and there are stars! Stars, so many, so bright, like you've never seen unless you went to the backwoods of Texas or somewhere. I don't know where the Moon is right now, I can't see it, but what I can see is our destination, the GeoStation, way in the distance. It is apparently half a kilometer across and under construction, but at this distance, with all the lights on, it's just a pinprick in the sky.

We aren't weightless yet. Apparently we've lost about 25% of our weight - I do feel lighter, it's easier to move, like wow, I must lose this weight when I get back to Earth! Back to Earth - now there's an expression I've never been able to use before. It dawns on me, I am somewhere else, other than on our planet! Like, wow!

So, even at 75% weight, we can still stand on the floors and climb the stairs. The upper floor has the bathrooms (always important to know where the bathroom is on a spaceship!) and dining area. So I'm up here now for dinner. Sad to say, it's microwaved, and tastes just as bad as airline food. I asked for vegan food. I am vegan - well, sometimes anyway, but it's a trick I learned when flying on an airplane to pre-order vegan food, it actually tastes better than the standard meal they serve.

April 12, 2033 2000 hrs

Bedtime, which means I have my PJ's on, my seat folded flat into a bed and I'm in it, exhausted after the excitement of the day. They'll finish transferring the water in the middle of the night sometime, so when I wake, we should be on our way. Goodnight all!

April 13, 2033 1200 hrs

April 13, a Wednesday, not an unlucky Friday, thankfully. Anyway, good morning people of Earth! This is your space correspondent Livi reporting from about 2,650 km above the surface of your planet!

We are well on our way, though the entire trip takes a week. I wonder how long before boredom sets in? I mean, a 24 hour trip on Earth is bad enough, but a week? But the good news is we lose weight rapidly, best diet I ever did haha. So now we only weigh half what we do at sea level. It's not weightless yet, but I can spring up and hit my head on the ceiling. Ok, so call me stupid, there must be safer ways of testing this.

It feels a bit like being on a yacht. The ribbon moves about, and we move with it, twisting a bit, and it feels like when waves move your yacht around, going up and down, a bit sideways, then back again. To be honest, I felt ill with it earlier on, what with sea sickness and space sickness from the low weight, my stomach is definitely complaining. They tell me it settles after a few days. Good news, not!

April 18, 2033 1400 hrs

Hi, me again. If you've been following my live vlog, you'll know we've settled into a routine, a boring one. The excitement of being in space gives way to boredom as one day after another passes, like being stuck in your house, snowed in, during winter, with nothing to do except read a book or watch a movie. A couple of times, we've passed downdraft cars heading back to Earth.

They tell me boredom is a good thing, in space. It means we haven't had any emergencies like solar flares, asteroids hitting us or whatever, so I should be grateful I guess.

We are now just about weightless, a fun thing. We entertain ourselves playing 3D pool with ourselves as the balls. Going to the bathroom now involves showers standing in tight water-capture capsules, while the toilet - let's just say its complicated.

But the exciting thing is, we're now only one day from the Geo-Station. Looking out the up window, we can see it, still small but more than just a bright dot at last. One more sleep :)

When we first started off, we had the feeling of climbing, of traveling up from Earth. But, now, with hardly any gravity to notice, it feels more like traveling along a railroad, sideways. My mind has sort of rotated everything 90° so I have a sense of Earth on my left, traveling to the GeoStation on my right, if you know what I mean.

So, in my mind, what was the "up" window is now the front or forward window, and I look along the ribbon towards our destination, as though it's a railroad heading for Grand Central.

April 19, 2033 1600 hrs

Hey. At last the excitement is ratcheting up again. We are an hour out from the GeoStation, about 200 km away. After looking like a distant dot for so long, it is growing rapidly in size every time I come back to look at it.

We are weightless now. I've changed into my trackies, apparently the favorite clothes up here, a tracksuit being practical as well as offering some warmth. I've packed my bag, not easy with no gravity, when things drift away from you, and as I discovered more than once, jumping up to catch something is not a good idea. You just keep going until you can find something to hold onto.

My stomach is still not great, but not as bad as it was. The movements of the ribbon, so unsettling further back, seem less here by the GeoStation.

April 19, 2033 1800 hrs

Hey crew, so here we are, finally on the GeoStation! Having arrived, it now looks big, so big after just looking like a dot all this time. Instead, it is Earth which looks small, looking back towards it, out of the window. Plus, I also got a view of the Moon, lining up with us, before I left the car. The Moon seemed like twice as big as seen from Earth.

The Arrivals Hall on the GeoStation: it's smaller than an airport one, after all, there are only half a dozen of us arriving. We take our own baggage along, pushing off and drifting towards Passport Control. The walls and things have handles everywhere, so you have something to catch hold of.

There's an African-American woman called Tamyra on Passport Control who gives me a broad smile, better treatment than any I've had back on Earth. "Welcome to space and to the GeoStation", she says, and means it. My passport gets stamped, and she hands me the first of important goodies I need up here - some slip on shoes which have a mild magnetic sole, sticking to the floor. Thank goodness, now I can walk on the floor!

Or not, as you like it. I see some people effortlessly flying from one door to another, amazing, but colliding with anyone is bad manners, I soon learn. GeoStation seems big at first, but it seems to shrink the longer you are here, until it feels like a hotel, split into rooms just the same, only every window you look out of, seems to have a view of, like, scaffolding, frames, things under construction. One day, they tell me, it will be many kilometers across, hundreds of times bigger.

I have a room of my own, for a night, just like a hotel, but otherwise it feels like I'm still on the cable car. I can float around the room just the same. There is a real dining room here, with a real kitchen, and though the food is still pre-packaged it does seem a bit better than we had on the journey. There's no wine, no alcohol, in space, according to the regulations, no smoking, no drugs or other bad habits. (What other bad habits? I wonder, have I missed out on anything?) I'm surprised someone hasn't tried smuggling some wine, at least, but it's dismissal and a return to Earth for anyone caught. Which leads me to wonder, caught by who? Can't see any police up here, all the people seem to be working on building the GeoStation. Then it dawns on me, everywhere there are cameras, no privacy up here, it would be impossible to get away with anything.

What does that mean, for, private moments, you know, like does anyone have sex up here? A discreet question over the dinner table brings laughs all round. Seems that is the first question anyone asks. The answer, they tell me, is yes, not that I'm looking to try it out, but apparently sex in zero-gravity is far more difficult. Plus, the delicate question of, sex with whom? There are no couples up here, the balance seems to be about 65/35 men/women and, so far as I gather, about 85/15 straight/gay, and no-one wants the reputation of being the first slut in space. So, while it has happened, on the whole it seems the space station is not the free-for-all some might imagine!

There are people here who have been to the Moon. I'm a bit wide-eyed about that, but they seem to be easy going about it, kind of like saying, oh yeah, I've been to Hawaii. But I'm like, hey, seriously, the Moon? It was half a century ago, the last time men (and I do mean men) went to the Moon, and that was a dangerous trip in an old tin can. Like, I'm going to be number 68 on the Moon, woman number 17, few enough of us, so it feels special to me. Ok, it's not number one, Neil Armstrong got that, but number 68 out of seven billion people! Hey, I don't care what you say (especially you Ginny) I feel special.

Cue James Brown, "I Feel Good" as I dance around in zero-g, bouncing off walls, floors and ceilings.

April 20, 2033 0900 hrs

Hey. After a one night break at the GeoStation, I'm in a cable car again. This time, we are heading further out from Earth, along the ribbon to the last station, 100,000 km from Earth, which everyone calls going "far away", to the Far Far Away Terminus, though its real name is the Outer Space Station.

It's a scary feeling, again, pulling away from the GeoStation. I know we've been in space for a week, but, weirdly, it now feels like we are really in space. Until now, everything seems to have centered around the Earth, leaving it, looking at it, but this is the first step into deep space. We really are going far away, leaving the planet.

At GeoStation, we were 35,500 km from the Earth. We're now traveling another 65,000 km or so away, so it really is far away. Less than a hundred people have ever been this far from Earth. That ought to deserve some champagne but like I said, no alcohol out here. We did get a tube of orange juice to celebrate, though.

Because apparent gravity is almost non-existent on this stretch of the ribbon, we go faster, about 400 kph, but it still takes another week to get to "far away". It doesn't feel like "up" so much as "out". I spend less time looking at Earth and more time looking at the stars, the sun when it comes around, and the Moon if it's in sight.

The apparent gravity is weak, even so, it's like the car has turned around, with the floor below us, while the roof looks back towards Earth. Now it's another strange feeling: gravity appears to start returning, but this time we are being pushed forward instead of backward, only mildly though. Our seats have been reoriented and I have the sensation of traveling backwards. In a sense, we are experiencing "negative gravity". In reality, the rotational velocity of the ribbon has exceeded the pull of Earth's gravity and we are being thrown away from Earth.

Now the car can travel faster. In fact it barely needs any power to move, since it is accelerating away from Earth also. Traction is maintained with the Elevator ribbon to steer it, but the next time power is required will be for braking and slowing in preparation for stopping at the end of the line.

April 27, 2033 1800 hrs

Hey again. A week later and we've arrived at "far away", the Far Far Away Terminus. My mental image has changed; now it feels like we are in deep space. The GeoStation felt like that when we arrived, but it was big, and the hotel feel was comforting. But the Terminus here is small and I feel fragile, way out in space.

If the GeoStation felt like a hotel, this place feels like a camper van. It's small, like the ISS used to be, I guess, with only three crew stationed here. It looks more industrial too, with tanks of fuel, air and things. The car we arrived in has left already, returning to GeoStation with a crew changeover. How ever did the old astronauts survive in the Apollo modules? It must have felt like hiding in an old tin can.

The other difference, at this station, compared with GeoStation, is the number of spaceships. This is the jumping off point for space travel, and a number of spaceships can latch onto the space station, being refueled and provisioned for their journey, though at the moment, there's only one out there, the one I'll be traveling in. This involves the movement of much fuel, which makes it a hazardous place to be. It is not a place where we want to be if an explosion occurs. That's why the harbor for the spaceships is separated a little way from the terminus. An umbilical ribbon, about one km in length, extends out from here, with service corridors and fuel lines, and at the end of the ribbon is the harbor where the spaceships are anchored.

They say there's gravity here due to the centrifugal force, but it's so weak you can't tell. It's just, if you're floating around, eventually you end up drifting towards the floor.

At the tiny docking station our next car awaits. No, it's not a car, it's a true spaceship, a rocket, to leave the ribbon and fly us to the Moon Ribbon. We have to travel from here, down the last bit of ribbon, in a small shuttle car, wearing our space suits, then transfer, via a docking corridor, into the space ship. Now it gets even scarier. I'm attached to the ribbon in more ways than one. Once in the space ship, we are trusting the rocket and AI computers to get us to the Moon. If something goes wrong, well, there are other space ships out here, but this is the riskiest part of the journey.

April 28, 2033 1100 hrs

So, hey, I'm in the spaceship and we are on our way. I can't see out; there's only a window near the nose, where the pilots are sitting. I have a screen to show the views but everything is so far away it isn't showing anything but stars. I have an ear-worm going, David Bowie singing Starman! Well, star woman to be precise.

We felt about 2-g of gravity when the rocket motors fired, after we had first nudged the spaceship away from the docking station, but now I feel weightless again and I've got the vomit-bag handy as my stomach doesn't like this. You'd have thought I was used to it by now. How many weeks have I been out here? Three, four, it seems an endless trip.

At least the spaceship bit only takes about 24 hours. How can people even think of traveling to Mars? It would take months and believe me, after the excitement of being in space wears off, boredom sets in. Now I know why they taught us how to meditate, to while away the time.

You are reading my blog and seeing us live on video, for what it's worth. There is wifi out here in space, with a herd of shoebox satellites carrying the wifi signal from anywhere around Earth, right out to the Moon, though I'm glad I'm not paying the Carrier bill!

Ok, back to the spaceship stuff. We left the ribbon, 100,000 km away from Earth. We are heading for the Moon ribbon. The Moon is about 400,000 km from Earth and the Moon ribbon is 200,000 km long, so the gap between the two ribbons is 100,000 km in a straight line. In practice, compared to where we were, back on the Earth ribbon, the Moon passes by about once every 25 hours, nearly. So our spaceship has to move out from Earth another 100,000 km, also tilt the trajectory to match that of the Moon, and time it so the Moon is swinging by, just as we reach the right spot to catch the Moon ribbon. So we take an elliptical trajectory, with the rocket motors kicking up our velocity when we left the Earth ribbon, and slowing us down to match the Moon ribbon when we arrive.

All this is done by the AI computer, and the pilots on board, and they assure me it is routine. All the same, it's a strange feeling floating in a spaceship, attached to nothing, and I'll be happy when we get there.

April 29, 2033 2300 hrs

Heya! It's been a busy few hours docking with the Moon ribbon and now we are in a cable car descending to the surface of the Moon!

That's the good news. The bad news is, even though we travel at close to 500 kph, it still takes us nearly a week to get there. This trip seems to be long periods of boredom punctuated by moments of high terror! Ok, I'm exaggerating but only a bit. In fact I get to look forward to the busy times, at least there is something going on.

Docking was the reverse of leaving. The Moon ribbon is 200,000 km long. The Moon keeps the same face towards Earth (you knew that, right?), so its ribbon hangs down towards Earth, through L1, with Earth gravity pulling it, so that the ribbon hangs in space. The other end is attached to the top of a mountain in a crater, Albategnius, which is more or less in the middle of the Moon, as viewed from Earth, and a stones throw from where Apollo 11 landed in 1969. (I wonder if it's literally a stones throw? I don't know how far a stone goes when you throw it on the Moon, I must ask someone.) On one of my days on the Moon, we'll get a trip out to visit the Apollo 11 reserve and see it for ourselves, wow!

Now we have proper windows again. This cable car is just like the one we left Earth in. I can see the Moon ahead of us (it's easier to think of it as "ahead" than "down"). What surprises me is how small it looks. I mean, it's a quarter the size of Earth, but from here, you realize just how small the Moon is. It looks kind of like a large ball and the curvature of the surface is really obvious.

So it feels, like, floating weightless again as we fall towards the Moon. With a good telescope back on Earth, you might just be able to spot our cable car.

Back on Earth. That expression haunts me now. I can see Earth behind us, it looks bigger than the Moon still, from here, but getting smaller by the day. Seeing the Moon ahead, it becomes real, the understanding that there are plenty of other places to go than Earth. We get so self-centered on Earth, like that is everything, but here I am, going to arrive at another place, our Moon. One day, it won't be just the Moon. We'll be able to travel to Mars and other places.

This is really hitting me right now. Space travel. It's real! After being just stories and movies, I am actually here. I've been on a spaceship and now I'm on an elevator descending to the Moon. I have to keep pinching myself to be sure it isn't a dream. It's like, wow, just, wow!

May 7, 2033 0200 hrs

At last, we've arrived! I'm standing on the Moon just like Neil Armstrong did! Me, Livi, reporting live from the Moon! It's taken four weeks to get here but now I'm so excited. I'm not on Earth, not floating in space, I'm actually standing on another place, our Moon!

Gravity is back, at least, it's only 17% of your gravity, back on Earth, and I can jump in the air and softly fall back to the ground.

Our cable car docked at the bottom of the ribbon, on top on the mountain. There is a docking station carved out of the top of the mountain, which gets enclosed with a shield once the car is inside, to form an airlock. They pump the cavity with air, and then we opened the door and could actually climb out, down a ramp and through another airlock into the arrivals lounge.

After being weeks in almost no gravity, my body suddenly feels like a lead weight, so heavy and I struggle to walk, even in such low gravity. But they assure me I'll adjust and get used to it.

First off, time to get my passport stamped, and yeah, I have an actual stamp in my passport saying "Albategnius Crater, US Territory, the Moon"! I've been told the colony actually has Starbucks and McDonalds here, small outlets but hey, I'm hanging out for something that doesn't taste like airline food.

This crater has the mountain in the middle of it, 12 km in height. We are at the top of it, and we work our way down. Inside the mountain they are tunneling and building a colony. It makes sense, protected from outside by the mountain itself. Accommodation levels are a good kilometer down, and there are shafts going all the way down to the crater floor level. One day, they'll make a huge CNT net, connecting the crater rim to the mountain, and cover the entire crater with, like, a huge umbrella canopy, fill it with air, and then we'll have a big, real, colony!

May 8, 2033 1300 hrs

So I found my room, had a nights sleep. Oh it's so good to have some gravity, you miss it. Now I'm in a viewing room on the side of the mountain. There is a large picture window and I can see out, into the crater. That the Moon is small is obvious, you can see how the surface curves. The crater walls are the horizon, I can't see anything past them, but even though they are only about 5 km away, you can see the land curving away.

Down on the crater floor, there is work going on. I can see what looks like tractors and trucks, lights and things. My first impression is the lack of color. The Moon is grey! Everything is just shades of grey. There's no air out there, so you need a spacesuit to be outside. That means no weather either, no clouds, no rain. The

Sun is bright, fiercely so, you wear, like, heavy duty sunshades. But look away from the Sun, and the sky is black, with the stars visible.

Up above me, there is a window so you can see up, the reason of course is to see Earth, which is always above us, here on the Moon. It looks small, now, far away. Now it feels like the Moon doesn't move, and it's the Earth that rotates just over once a day. The Earth seems to move into light and shadow, day and night, too, but that's just the illusion of how it looks from the Moon.

It's my first day here, having taken a month to get here, but I can understand the appeal of living on the Moon. It might as well be another planet, as far as we are concerned, "we" being the inhabitants of Moon. (First thing I learned, don't call it "the Moon", here the trend is, we just call it "Moon".)

Lunar colony concept

The Earth, well, you know how, back on Earth, you look up and see Moon? Well, here, that's how people think of Earth. They look up and see Earth, the only difference being Earth is always there, hovering in the sky.

But, for the colony, most of life goes on inside the mountain. Once inside, it's like living in a super large building, with levels, stairs, lifts. If it wasn't for the low gravity, you could forget being on Moon.

Going on the tour to see the Apollo 11 command module will be something special. I'm like, wow, it really hits me, I am standing on Moon, the elevator does work, space travel is real.

And this is the start. We'll build more colonies on Moon, then we'll do the same on Mars and other places. It's like I traveled with Columbus, seeing America for the first time, I know how it must have felt. For we aren't bound to one planet anymore. We've got a whole solar system to explore. Was it worth it? You bet! A month-long journey can seem like hell, but when you get out here, it's worth it.

THE SPACE
ELEVATOR SERIES

BOOK ONE: THE SPACE ELEVATOR CONCEPT

Published 2020

Linda J. Phillips

The Space Elevator 2020 series
Book One:
The Space Elevator Concept

Publisher contact: info@21stcentury.space
FaceBook www.facebook.com/lindyjaniceAuthor
Twitter @_lindaphillips
Amazon author page https://www.amazon.com/author/linda-janicephillips
Web 21stcentury.space

Web links utilized in this publication were correct at the time of writing, but they can change over time.

Published by Linda Phillips

www.ingramcontent.com/pod-product-compliance
Lightning Source LLC
Chambersburg PA
CBHW070410220526
45467CB00001B/522